To Sy Krim —

old and dear friend
not for reading — just
a memento

Lee Wiggins

PANEL ANALYSIS

Progress in
Mathematical Social Sciences

PANEL ANALYSIS

Latent Probability Models for Attitude and Behavior Processes

LEE M. WIGGINS

Elsevier Scientific Publishing Company
Amsterdam · London · New York 1973

For the U.S.A. and Canada:
JOSSEY-BASS INC. PUBLISHERS
615 MONTGOMERY STREET
SAN FRANCISCO, CALIF. 94111

For all other areas:
ELSEVIER SCIENTIFIC PUBLISHING COMPANY
335 JAN VAN GALENSTRAAT
P.O. BOX 1270, AMSTERDAM, THE NETHERLANDS

For the permission to republish the two lines quoted from the poem "Among School Children" from "Collected Poems" of W.B. Yeats grateful acknowledgement is made to The Macmillan Company and M.B. Yeats and Anne Yeats.

For permission to republish the line quoted from "Hop" from the "Collected Poems" of Thomas Hardy grateful acknowledgement is made to The Macmillan Company and the Trustees of the Hardy Estate.

Library of Congress Card Number: 72-97441

ISBN 0-444-41112-7

Printed in The Netherlands

Progress in Mathematical Social Sciences

Other books included in this series

Raymond Boudon
Mathematical Structures of Social Mobility

J.K. Lindsey
Inferences from Sociological Survey Data – A Unified Approach

Abram DeSwaan
Coalition Theories and Cabinet Formations – A Study of Formal Theories of Coalition Formation Applied to Nine European Parliaments After 1918

Contents

To Paul F. Lazarsfeld and Ernest Nagel

Preface

No one knows how many billions of items of panel data have been collected in the United States. It may well be that the number of such items collected each year runs into billions. By "panel data" is meant observations of behavior — physical, oral, or written — of a set of individuals, families, households, stores, firms, or organizations at two or more points in time. Most of the data, and most of the attention here, refers to the individual person, but, mutatis mutandis, similar concepts and models can be used for the larger units of analysis.

The largest collector is, of course, the Federal Government, with the Census Bureau perhaps in the lead, partly because it conducts studies for other agencies than itself, such as the Federal Reserve Board. There is the vast work of the Bureau of Labor Statistics. The Social Security Administration has continuing files on millions of participants, as does the Internal Revenue Service and the Pentagon. The whole educational system, which is mainly local and state-wide in its administration, has continuing records on scores of millions of students running over a period of years. There are vast studies of housing, moving, re-location, and so on. There are also draft boards, security and police systems. There is the enormous apparatus of civil service on all levels, federal, state, and local, with personnel files, pension systems, and the like. There are panels of farmers estimating crops. Some of the government records and interview panels contain hundreds of items on each person running over many years.

The private sector is not far behind. Company personnel files cover scores of millions of workers over many years. Many tax, social security, and pension files overlap government data, but by no means all. In the field of consumer research alone it has been estimated that upwards of three quarters of a million diaries, usually filled out daily or weekly and each involving hundreds of items, are collected annually by commercial research firms. Some of the participants have belonged to panels for twenty or more years. There are mechanical devices attached to television sets which give virtually continuous readings, running for years. Media research panels collect vast amounts of panel data on viewing, listening, reading, advertising exposure and related variables. There are large-scale panels of economic anticipations, consumer behavior, brand selection, and so on, conducted by commercial firms, industrial firms, financial institutions, non-profit consumer organizations, advertising agencies, trade organizations, and others. Foundations conduct huge panels on family planning, health services, religious attitudes. Polling organizations run thousands of panels on candidates, issues, attitudes, etc., in every election campaign.

In addition, there is a myriad of smaller-scale panels — evaluations, quasi-experiments, straightforward observational studies over time in classrooms, small organizations, small communities, dealing with a great variety of behaviors and attitudes.

Most of this rich lode of panel data is never analyzed in terms of change, other than on the simplest aggregative level, and much of the analysis which is done is incorrect. The panel is ordinarily used mainly as a convenience; for example, it is cheaper to re-use a sample than to draw a new sample, in most applications. It makes possible the cumulation of more data on a given person or unit than could be done in one interview or questionnaire. It makes possible certain verification and clean-up procedures.

These are useful and important functions, but it is arguable that what is thrown away or unused is as valuable as what is used. Indeed, the study of movements of a variable over time in terms of an internal rather than merely aggregate analysis is often necessary in order to understand what is happening at one point in time.

This book deals with such movements, and in so doing cuts across a number of fields of application and a number of aspects of the study of behavior and attitude. Regardless of detail, those who attempt to interpret panel data, whether their field is commercial, educational, financial, welfare, politics, or in fact almost any field in which individual units are subject to change, will ignore the basic theme of this book at their own peril. That theme, most broadly stated, is that in dealing with panel data things are not what they seem, and that to take manifest panel data at face value is almost invariably to err greatly in interpretation.

This theme, in the present work, is elaborated in four directions, for it cuts across the usual classifications of existing fields. It is thus of importance to a great many occupational specialties. The four directions are not necessarily of equal or even any interest to groups for whom some of them may be all-important. They may be distinguished as follows:

(1) Methodological.

It is doubtful whether more mistakes of interpretation have been made in any design used in the social sciences than in that of the panel. These are both qualitative, on a simple plane of analysis which takes everything at face value, and highly sophisticated, applying a piece of existing mathematics which in itself may be quite complex and interesting to situations where the results are totally misleading. The present work began with data and anomalies found in close study of large bodies of data, which helped in avoiding some of the more obvious errors.

(2) Mathematical.

Once it became clear that the movement of a panel variable over time is a complex and composite process, a set of models was developed which differed from any previous models, and which had explanatory power going beyond any that existed at the inception. My first quasi-publication which was available on request to the interested reader appeared in 1951, but the work had begun in 1947, so a quarter of a century is reflected in what follows. Of course, my own models have been developed and my thinking has changed over that time, but for certain situations the original models are still applicable. Because at the time I began there were perhaps not more than a half dozen sociologists in the United States interested in mathematical models, I kept the mathematics on a simple level, and the non-mathematician can skip the algorithms without great loss. In this presentation I have decided to keep the mathematics on that level.

(3) Empirical.

Unless these methodological and mathematical models were applicable to a wide range of empirical panel data, they would be of no interest to me, and of only limited interest to anyone else. As I stated, I started with data and tried to find models that would fit. I did not fit every body of data I analyzed, but I did fit a great many, from an intuitive point of view (that is, I do not employ statistical tests of goodness of fit). A noted mathematician who heard me present some of these models in a faculty seminar complained that I had too many models. But life presents us with a great variety of situations, and many of these crop up in the kinds of panel data we collect. These models cut across many empirical fields of investigation, and they require appropriate modifications accordingly. If they did not, I myself would suspect that something was wrong. Human behavior, response, attitude, is not that simple. The model which is going to be appropriate for the study of planning to change residence is not the same as the one which is appropriate for studying changes in educational performance; but there are great underlying similarities.

(4) Theoretical.

Any work which is primarily based on the vast realm of usage in a number of empirical fields can draw from many fields of theory, but on a rather broad or abstract level. Yet, although many readers may not be concerned with social or behavioral theory, there are important underpinnings to these models in social theory. Volumes could be written on this score, but for this monograph the theoretical considerations have been compressed. Pareto, Kurt Lewin, and some of the more recent work on cognition have been chosen for their relevance. I have also invoked a few concepts of my own. For some of the models, Pareto's theory of sentiments, residues, and derivations has been most useful; for others, the concepts involved in level of aspiration, levels of reality, and so on, have been valuable. All of this could use more development than it has been given here.

In sum, the reader will find a number of mathematical models applicable to panel data, with a number of empirical applications and examples given and some methodological implications noted, together with selected theoretical underpinnings. Coming from a background in philosophy, with my first publication (1941) in that field, I have tried to control the temptation to move in that direction more than minimally, since this work is aimed chiefly at the enormous class of those who work with panel data in one way or another.

The principal aim here is to resolve panel univariate data into components which reflect quite different aspects of human behavior or attitude. As far as I know, my original publication (1951) in the Columbia—RAND project directed by Paul Lazarsfeld is the first work of this sort. The only monograph of which I am aware which tackles the same problem (and some others) head-on is James S. Coleman's *Models of Change and Response Uncertainty* (1964), which develops a different set of models for the same resolution, using continuous-time rather than discrete-time models. These models are to be distinguished from learning models, of which the leading modern example, the Bush—Mosteller model, was developed at about the time I began my work.

It is a great pity that the non-learning models have not been taken more seriously by both academic persons and practitioners of the panel technique. Many errors could have been avoided, some of which have tended to discredit the use of the non-experimental panel. My work was generally available, but appeared in rather obscure places. The same thing cannot be said of Coleman's, which appeared in monographic form. His work is mathematically more difficult; on the other hand, he presents explicit computer programs for solutions.

It is true that no one conceptualization or set of models can be applied to all non-experimental panel data, and the routine application of concepts such as are described here or in other models addressing the same problems will require a great deal of work, based on the realization of the necessity of doing the work. Ignoring a problem does not make it disappear. It is encouraging that in the past ten years or so there has been a great increase in interest in the analysis of panel data, addressed mainly to the relationships of a system of variables moving over time. True, there is a persistent tendency to talk of "error of measurement" or "reliability" rather than to accept the basic probabilistic relationship between most observed behaviors and attitudes on the one hand, as measured in one or another way, and the underlying characteristics which give rise to them, whose metric may be totally different, and for which in any event the concept of "error of measurement" is largely inappropriate. Nevertheless, the increased interest in the kind of problem presented will lead to more meaningful interpretations, certainly in the huge area of practice, and eventually in the area of theory.

We social scientists have perhaps tended to overreach, regarding very complex phenomena as much simpler than they are, making unwarranted claims, borrowing from simpler fields such as physics and biology, and so on. But science is the method of self-correction. With the computer we can handle larger bodies of data, and with the computer we can solve the more complex equations of the more powerful and complex mathematics available today.

No practitioner or theoretician or methodologist of panel analysis can afford to be unaware of some of the concepts presented in this monograph. Three persons deserve mention in this Preface. Paul Lazarsfeld first gave me the basic notion that the manifest and latent levels do not have to have the same metric, or the same distribution for any particular sample or population. Chapter 9 of Lazarsfeld and Henry's *Latent Structure Analysis* (1968) is an augmentation of my earlier work, and Lazarsfeld has written several articles with variants of these models. James S. Coleman was kind enough to submit written suggestions for improvements on an earlier version of this monograph, which at one time was supposed to appear in a series of which he was an editor, but which was discontinued before the work was completed. And the general editor of this series, Raymond Boudon, has referred to my earlier work in one of his own monographs. *L'Analyse Mathematique des Faits Sociaux* (1967) and has also been most kind to me in the preparation of this text under difficult circumstances.

I gratefully acknowledge the following sources of support and help not mentioned above: Financial support — the RAND Corporation and the Social Sciences Division of the National Science Foundation under Project No. NSF—GS—992. Fellowship support — The Center for Advanced Study in the Behavioral Sciences and the Italian Fulbright Commission. Released time — the Graduate Sociology Department and the School of Social Work of Columbia University. Help with suggestions and data — T.W. Anderson, Nelson Foote, Albert Hart, Herbert H. Hyman, F. Thomas Juster, Patricia Kendall Lazarsfeld, Bernard Levenson, William McPhee, Robert K. Merton, and Belle Wiggins.

CHAPTER 1

Introduction

Plato thought nature but a spume that plays
Upon a ghostly paradigm of things
William Butler Yeats

This monograph deals principally with the structure of a single item of human behavior (including in particular verbal behavior), moving over time, usually in non-experimental situations. If that sounds simple, the reader is mistaken. The law of falling bodies deals with an even simpler situation, since inanimate matter is simpler than the human organism, yet perhaps the greatest thinker of all time, Aristotle, who incidentally spent some time on this subject, did not discover the law. That discovery was left for a great but lesser man, Galileo, whose work was then used by arguably the greatest scientist of history, Newton, who devoted considerable thought to that same seemingly simple topic.

There was a myth about this matter which was propounded by many teachers of philosophy to many students for many years, which went something like this: Aristotle was a theorist with little interest in close observation of nature. Without bothering to confirm it carefully experimentally, he theorized that a heavy object would fall faster than a light object of similar shape and volume. He was followed by centuries of philosophers, especially Christian philosophers of the "dark ages", who were equally uninterested in close observation, and besides, were not much interested in the local movement of an inanimate body falling on the earth's surface. Suddenly, along came Galileo, who observed falling bodies very carefully. He went to the top of the leaning tower of Pisa and dropped two objects of equal shape and volume, one heavy, one light. Measuring their fall closely, he found that they reached the earth simultaneously, thus giving rise to the notions which were required for the law of falling bodies.

Today every schoolchild knows that all this is nonsense. Aristotle was a close observer (of course he made some mistakes), and so were many of the Christian and other physicists who preceded Galileo. If the leaning tower of Pisa experiment had been conducted, the heavy ball would hit the ground before the light ball. Why? Air resistance, which would slow down the light object more than the heavy.

The genius of Galileo here was to conceive something that did not happen concretely, but would have happened if there had been no atmosphere, as in a vacuum. From the point of view of the law of falling bodies, the atmosphere is garbage, and it required several thousand years to clear that garbage out of the minds of the scientists. But one man's garbage is another man's meat. Some of the tastiest barbecue ever eaten was fed on the "edible garbage" sold to growers from military mess halls — the label was on the garbage can when the troops went out of the hall. To the Wright brothers and other engineers of propellor-driven heavier-than-air craft, or for that matter, even to space scientists, the atmosphere has many useful properties, including the fact that people like to breathe.

Indeed, either the disposal or the recycling of garbage has provided notable advances in science. Sir Alexander Fleming took other research scientists' garbage and made penicillin from it. He is reported to have remarked, on seeing the relatively sanitary conditions of biochemical research laboratories on a trip to the United States after World War II, that if the British laboratories had been that sanitary before the war when he was working in them, penicillin would probably never have been discovered.

A great deal of effort in the social sciences has been devoted, quite properly, to cleaning up the data. In psychological testing, in econometrics, in survey research, in the current interest in social indicators, and so on, a great deal of effort has been expended to produce measurements with less error than was previously hidden. Many substantive studies have been exploded by the discovery of hidden errors. Donald Campbell, among educational psychologists now working, is exceptionally adept at this sort of work (see for example, Campbell and Stanley, 1961). McNemar made much of his reputation in exposing fallacies caused by regression on the mean, among other errors.

McNemar (1946) wrote an important review article on survey research, in which he pointed out that although psychologists had done a great deal of work on unreliability in the field of testing, survey research workers had largely ignored the problem, depending without qualms on single items. Shortly after that time I was asked by Paul Lazarsfeld to re-analyze some of the panel data from his 1940 voting study (Lazarsfeld *et al.*, 1948). I was also asked to try to locate data from the files of the Bureau of Applied Social Research at Columbia (including the study mentioned) which might fit the Markov chain models then being developed by T.W. Anderson (1951, 1954), among the first models suggested for the movement of a single item over time.

This investigation led me rapidly to empirical findings of a peculiar nature, which led to the development of a series of models, some of which are included here. A paper which I wrote was included in the same project in which the Anderson paper was included (Wiggins, 1951), and was the first publicly available statement of the approach taken here. Subsequent to that time, my ideas have undergone some modifications, partly as a result of the examination of more data, partly based on subsequent familiarity with latent structure analysis and partly on my own reflections.

The problem having been raised by McNemar, I began to look for ways to determine the reliability of a single item, as used in surveys, and in particular in repeated interviews with a sample of identical persons — that is, so-called "panel" studies. The type of curious finding which one makes with some variation or another is that the amount of change of position on a great many items depends to a very slight degree on the time interval. If we are talking, say, about a three-interview panel, we often get results like the following (Wiggins, 1951, and 1955a).

		Time 1	
		yes	no
Time 2	yes	A	B
	no	C	D

		Time 1	
		yes	no
Time 3	yes	$A-\alpha$	$B+\alpha$
	no	$C+\alpha$	$D-\alpha$

		Time 2	
		yes	no
Time 3	yes	$A+\beta$	$B-\beta$
	no	$C-\beta$	$D+\beta$

Here, A, B, C, and D are numbers of cases, with B and C showing what is often called the "turnover", and the relative size of B and C depending on whether and how the marginals are changing. The Greek letters α and β refer to numbers which are usually positive but relatively quite small and often negligible. The result is that the turnover between time 1 and time 3 is usually slightly larger, but not much larger, than the turnover between time 1 and time 2, while the turnover between time 2 and time 3 is usually slightly smaller than that between time 1 and time 2. The situation schematized here is one in which the marginals do not change between waves 2 and 3, but comparable results are often found, mutatis mutandis, for situations in which those marginals do change.

The question whether α and β are equal to zero, or whether either is equal to zero, or whether one is larger than the other, is one of the matters which we shall want to discuss in the concluding chapter, since it depends on the structure of the situation, of the item, on whatever theory can be brought to bear, and many other matters. Here, we merely wish to review some of the major methodological and formal considerations which enter into the development of models which may describe some of these findings. This involves some attempts at classifying the possible structures in terms of the measurement space and time, as well as indicating some of the assumptions and limitations of those structures with which we deal.

Categorization by Measurement Space, Time, and Systematicity

Evidently, there are limitless numbers of ways of categorizing the types of data which arise in looking at one variable used in non-experimental or quasi-experimental studies of a sample or population over time. These occur on both the manifest and latent levels, which will be defined below. Let us review some of these possibilities in terms of the type of measurement, the time scale, and the systematicity of movement of the variable.

(1) Measurement

There are many excellent discussions of types of measurements (see for example, Torgerson, 1958, and Blalock, 1960), so that no attempt will be made here to discuss the various possibilities in detail. Among them are the dichotomous attribute, the set of non-ordered categories more than two in number (as ethnicity, for example), the set of ordered categories (as income groupings), a complete ranking (rarely used with large samples), the ratio scale, the interval scale, and so on. But in social research we also often use multiple responses from one subject at a given time, which means that each item may have a varying number of different responses for each subject, which are not mutually exclusive or independent.

Again, we often combine responses in various ways through scaling techniques, so that the resulting manifest indicator may be an index with several ordered or unordered categories or even an approximation to a continuous measurement, even though it is built up from dichotomous items. Methods such as those for constructing psychological tests, various forms of scaling, latent structure analysis, factor analysis, and so on, yield measurement constructs of many kinds.

Coding of open-ended questions, or content analysis of clinical documents (and there is a vast literature on these subjects), may lead to very complicated types of measurement. It is scarcely comparable to measuring the length of a room with a ruler.

We shall presently discuss the latent—manifest relationship briefly, but it is sufficient here to say that the latent measurement space may be as complicated as the manifest space, and similar to or different in form from the manifest space.

To complicate matters further, we often find that the manifest measurement is changed over time. Many research persons conducting panels or longitudinal observations of the same subjects over time change the question or response wording from one wave to another, sometimes for good reason, sometimes not. Sometimes the panel director has little interest in changes over time, our major concern here, and merely wishes to cumulate data on the subject, or to verify the data or increase the reliability. Often judges are used to categorize responses, and their criteria may change with time.

(2) Time

While clock or calendar time may be taken as continuous, it is a rare non-experimental study which will have a continuous measurement of any variable over time. Some of the items studied may be thought of as continuous over time, even though the measurements are not. An example might be one's attitude towards a political figure. Even there, a given respondent may at a given time have no attitude on a given issue. Further, the time interval at which one asks the respondent about the public figure may be constant or varying. Some items have a built-in regularity, such as publication of certain media, appearance of television programs, etc., and in those cases it is customary to employ a constant interval for obtaining information, although the information does not necessarily include every occurrence of the publication or program.

Sometimes the time is measured by an event, rather than a period of calendar or sidereal time. Thus in market research, especially on small items, it is not uncommon to use consumer diaries, which are filled in when purchases are made. Studies of brand consumption may not distinguish between the frequent buyer and the infrequent buyer, so that the movements from one brand to another for a given purchase may represent for one respondent a time period which is much shorter than for another. So-called "repairmen" problems in probability often have this character (Feller, 1957 and Barlow *et al.*, 1965). On the other hand, interest may center in the number of purchases made over a certain calendar period, so that the measurement is a cardinal number taken during a constant time interval. Or a young woman may be asked how many children she intends to have.

(3) Systematicity

There may or may not be any systematic movement of a single variable (or a group of variables) over any given time or any given sample or population. Of course, it is one aim of developing mathematical models in this area to locate and describe such regularities. Given the enormous complexity of the situation, one would be happy to find any area of regularity with a model to describe it completely.

However, the structure of the situations studied is itself subject, in most areas, to such broad changes, that we are likely to catch any systematic movement only fleetingly, if at all. There are certain broad classifications into which we may place many of the situations which we encounter, and there is some qualitative theory to support their allocation. This will be discussed further in the concluding chapter.

But just as the models proposed in this monograph were developed primarily inductively from looking at hundreds of sets of panel data from scores of panels, and observing several types of qualitative regularities, so it is likely that a great deal of further study of data will be required to develop models which completely specify any of these situations. However, a great deal can be learned about the movement of a single item without necessarily specifying every property of its movement, information without which we could not hope ever to arrive at completely specified models. Important regularities may be observed, important parameters determined, without knowing everything.

The idealist logician Bradley took the view that every concrete occurrence was related to every other concrete occurrence, and that all these relationships were constitutive of the occurrence, and that therefore in order to know any one thing one needed to know everything about its relationships with everything else in the universe. Since one could not know that, one could not know any one thing, and therefore one could not know anything.

Bertrand Russell is reported to have said with characteristic wit and cynicism that he had spent a good portion of his life in the study of logic and made many important discoveries, of which the most important was that the whole subject was trivial.

Still, science and technology inch on, for better or worse, and men have stood on the moon. We must walk the tight rope between despair and over-optimism.

The Manifest—Latent Relationship

Science commonly deals with non-observables, as noted in reference to the law of falling bodies. For example, while the concept of "weight" may be said to be observable in natural conditions on the surface of the earth, the concept of "mass" is not. This is true, of course, for human behavior, including verbal behavior, when it is analyzed closely. Although many human acts are directly observable, many are not, and the possibilities of manipulation of behavior, especially the verbal, are so great as to make it seem likely that the study of non-observable properties will loom even greater in the behavioral sciences than in the physical sciences.

The "manifest—latent" terminology has been employed in so many ways and for so long that one might wish to avoid it. But the sense in which I wish to use it corresponds so closely to the sense in which Lazarsfeld uses it, although it is not identical, that it would be straining to use another set of terms. The first chapter of Lazarsfeld and Henry (1968), as well as many other publications of Lazarsfeld, describe his usage so explicitly that I shall not go into detail; in fact, the sequel of this work will make amply clear the usage.

Some of Lazarsfeld's concepts and related notions are discussed briefly in Chapter 2 below, but in relation to the citation above, there are three points to be made here.

First, I do not find Carnap's (1936, 1937) definition of a "disposition concept" to be particularly useful for my purpose. Latent structure analysis is a measurement model, designed to elicit a latent concept from a number of manifest items, taken at one point in time. The models here merely relate a single manifest item, taken over time, to whatever latent concept that item may imply. If the latent concept to which the manifest item refers changes over time, the model has no significance. As initially stated by Carnap the relationship was not probabilistic, but "if and only if" one thing happened then another happened a concept was partially defined.

Second, most important is the probabilistic relationship between the manifest and the latent. Here other terminology has been invoked, especially in test psychology, where the notion of "true" scores and "observed" scores has been used, the relationship being called "reliability." Having come into the subject from examining panel data on the basis of McNemar's paper, and having found a number of panel items in which the turnover from time 1 to time 3 was equal to the turnover from time 1 to time 2, I adopted that terminology, which indeed was used in the title of my first paper on the subject (1951). But the concept of "error" is taken from measurement theory, and I had already begun to realize its inadequacy for the type of data which which I was dealing.

The normal distribution often used to be called the "normal curve of error" because it was found to describe errors of observation made, for example, by astronomers observing a fixed star. In an elementary physics laboratory, students are often given the assignment of weighing the same object on a fairly sensitive scale some large number of times, recording the average position at which the needle comes approximately to rest each time. These measurements will assume the normal curve, approximately. Thus the normal distribution is often a good model for error distribution. The notion of reliability of measurement makes sense.

This was carried over into test psychology, although with some difficulty, because it was found that the reliability of a test, say, often differed for different groups of subjects. In other words, the same test often has one reliability for eighth grade students, another for ninth grade students, and so on. The difficulty is that the trait supposedly being measured is itself imputed. The comparability to a fixed star is tenuous, at best. The concept of "error" thus becomes a bit fuzzy. People have spent lives trying to define what the MMPI test measures. The situation with the TAT or Rorschach is even worse, or any so-called "projective" test. This is not to say that these tests do not measure anything, or have no value, but merely that what they measure is imputed, non-observable.

Thus although I have occasionally used the terminology of test psychology in this monograph, for convenient reference where I discuss continuous space variables, I believe that Lazarsfeld is right in rejecting the use of the concept of error and instead referring to "latent probability" as the basic term for expressing the relationship between the latent position and the manifest.

This term implies a basic probabilistic relationship between the latent position and the manifest expression. Of course, that probability could approach one, as in the case of a physical measurement of, say, a person's height. Further, error of measurement is compounded in practice with the latent probability; mathematically, they are inseparable. As stated above, there are the two approaches to garbage — one is to dispose of it, in this case to try to eliminate errors of measurement and indeed all probabilistic elements in a measurement, an effort which makes great sense in many census and economic measurements — and the other, to accept that there is an inherently probabilistic element in the situation and to try to "re-cycle" or make use of it in one way or another. The approach here is primarily the latter.

A few words may be added here about the concept of a probabilistic relationship between a manifest position and a latent behavior or verbal item. As noted below in Chapter 2, the notion of probability in no way implies any philosophical view of indeterminacy. The very fact that the level of probability itself may change, and that the probability may approach one, implies that fact. It is a question of the level of explanation. Certainly there is a degree of randomness in human behavior, taken on any given level of explanation. The same thing is true of error of measurement. We do not need to get into the question of whether God is playing dice with the universe.

Because it has become increasingly clear to students of human behavior over the past twenty-five years that the randomness found is not merely an error of measurement, some research workers have adopted the term "uncertainty" for this type of behavior or response. But having haggled about Heisenberg's uncertainty principle many years prior to getting into the present problem (see Wiggins, 1941), a principle which has a very specific meaning quite different from that which we refer to here, I do not advocate the use of a term which is already well-established in another context.

The third point I mention from Lazarsfeld's approach, and it is one which cannot be overstressed, is that there is absolutely no direct relationship between the form of the distribution on the latent and the manifest levels. Consider a characteristic such as "arithmetic ability." Without necessarily committing oneself to the view that for a given population this distribution is normal, it is fair to assume that it is approximately continuous. Now suppose that we have a test of two hundred items designed to test "arithmetic ability", with each item scored as "correct" or "incorrect". Each item is a dichotomy, and let us suppose that all errors of coding and so on have been eliminated. Any one item is a measure of arithmetic ability; thus we have a manifest dichotomy measuring a continuous variable. Obviously, the relationship is probabilistic. What test psychologists have generally done, of course, is to use a fairly large number of such items varying in difficulty, and summed them to arrive at a score which, when corrected for guessing and often standardized in various ways or even normalized, produced a distribution which they thought corresponded to the latent trait "arithmetic ability." Even then, they have often gone on to explore through factor analysis and other methods the dimensionality of the traits underlying a set of tests.

The point here, however, is that there is absolutely no inherent relationship between the type of measurement of the test item and the trait being measured. In the example given, a dichotomous item is used to tap a continuous latent variable. Another kind of example may be given from economics anticipations research. A respondent is asked a question like "Do you intend to buy an electric refrigerator within the next twelve months?" The response categories provided are "yes" and "no". Juster (1964) has concluded that each of these responses really encompasses a probability distribution of purchase. The purchase "intenders", those who say "yes", have one probability distribution for purchase, with a higher mean than the different probability distribution for the "non-intenders", those who say "no", but with considerable overlap in the distributions.

The converse is less often encountered, but it is entirely possible. It is most likely to be found in self-administered tests or forms where graphic rating scales can be used. For example, in Appendix C of this monograph, I show certain graphic rating scales with continuous lines running from, say, zero to one hundred, on which I asked graduate sociology students at Columbia University to mark their aspirations and expectations in various areas. Again, in Chapter 6 below I refer to a study of a large number of low-income parents, in which graphic thermometers were used for many items.

On such continuous response spaces, denoted by a single continuous line running from some "zero" to a "highest" point, or some version of that, one may of course obtain any univariate single-response distribution one can imagine. It is not uncommon to obtain what amounts to a set of responses which are very close to a dichotomous attribute, a piling up of all (or nearly all) at the end points (or very near the end points).

Again, one may ask respondents to mark on such a scale the answer to the question "Are you male or female?" and get what amounts to a continuous distribution (bi-model, to be sure). These responses are not necessarily flippant. Or one may ask, "For which party do you intend to vote?", assuming there are only two candidates who have already been nominated in a given campaign. Since there are only two candidates for whom one can vote, the fact that you may get a continuous although bi-modal distribution in manifest scores implies a lack of congruence between the manifest and latent measurement spaces.

As noted above, the manifest form of responses is often changed by those conducting panels. It is also true that the latent measurement space may change over time. Thus, for example, in a political primary campaign with four candidates, one may drop out, thus leaving only three possibilities. Again, in market research, a new brand may be introduced between panel waves, thus perhaps changing the possible latent classes, as well as the manifest.

The point is, to repeat, that there is no necessary and inherent correspondence between the measurement space on the manifest and latent levels. The limitations of data and the desirability of simplicity and parsimony have led us in this monograph to place most emphasis on models which display such correspondence, but it should be kept in mind that Occam's razor can cut muscle as well as fat if pushed too hard.

Classification of the Models of this Monograph

The two principal types of parameters of the models of this study are the proportions of respondents falling into each of a given number of inferred latent classes, and the probabilities for those in each latent class that their responses, behavioral (including verbal), will fall into any given manifest class. Either or both of these probabilities or proportions may or may not change from one time to another, and such changes may occur in systematic or unsystematic ways. Now, any systematic process can occur on the latent level, in one situation or another, and I have chosen to comment only on the latent Markov chain with a discrete latent and manifest space and a discrete time. The introduction of an explicit latent probability process of some sort can obviously reduce or increase the number of unknowns, while bringing into play parameters of various kinds.

The principal basis of classification I have used is to distinguish models on whether the latent probability for a given class is considered to remain the same (which is systematic at the lowest and simplest level — namely, nothing changes) and whether any respondent may be considered to have changed his latent position or not (and again, if no one changes, we have the very simplest type of systematicity or process). Thus for each of these two fundamental parameters we may have either no change, which is a very simple type of order, or we may have systematic change, as in a Markov chain, or we may have unsystematic change.

To anticipate, I may say that my best results have been obtained with the first and third of these situations. The third type does not imply lack of regularity, but simply lack of a mathematical model to describe it with any precision. In fact, regularities have been found, which will be discussed in the concluding chapter, and they are of considerable interest to the social scientist.

The three-fold categorization which I have applied to each of two types of parameters leads to a nine-fold classification of models on this basis alone. I want to give a possible physical analogy to each of these categories as I list them, following which I shall briefly discuss the question of discrete and continuous time models.

We may first outline the nine possibilities as follows:

	Latent probabilities over time	*Latent class positions over time*
(1)	No change	No change
(2)	No change	Systematic change
(3)	No change	Unsystematic change
(4)	Systematic change	No change
(5)	Systematic change	Systematic change
(6)	Systematic change	Unsystematic change
(7)	Unsystematic change	No change
(8)	Unsystematic change	Systematic change
(9)	Unsystematic change	Unsystematic change

Class 1

If we were dealing solely with errors of measurement, without any questions of memory, instrument effect (see, for example, Glock, 1952) etc., then the first model would be appropriate. As a matter of empirical observation, it comes a lot closer to many examples than one might expect. It would yield a value of zero for all the Greek letters in the paradigm of two-wave data given above, and the quantity B would equal the quantity C; that is the marginals would not change. For a physical analogy, we might say that we had taken a still photograph with a certain degree of resolution, not perfect but unchanging from one time to another, of an unchanging object under identical conditions, at three points in time. The three pictures would all be developed to the identical degree of contrast, all with the same degree of fuzziness, small or great as it might be. The lack of resolution or contrast would make things appear a bit different from one time to the next, even though they had not in fact changed. The fuzziness roughly corresponds to a latent probability less than one, and while the object is unchanged it appears to change somewhat because of the lack of resolution although the degree of resolution itself does not change. Frankel (1961) has developed some very interesting models of cumulative readership which fall into this group, based on the concept of latent behavior function.

Class 2

In the second class of models, some respondents change their latent positions, in accordance with some systematic process.

Chapters 2 and 3 deal with models of classes 1 and 3. Chapter 4 deals with a particular model of class 2 and latent Markov chain. Chapter 9 of Lazarsfeld and Henry (1968) extends this kind of analysis somewhat, and Lazarsfeld has proposed other models which fall into this class, which he calls models of oscillation and corrosion (see also Kendall, 1954).

Clearly, any type of process model can occur on the latent level, and in general if it does it will not occur under the same guise on the manifest level. The analogy would be to a series of still photographs developed and produced identically, made at a series of time points where the object being photographed was undergoing a systematic change. For example, we might be studying an experimentally controlled situation in which a sample of subjects is being moved in a given direction linearly by a controlled stimulus. Or, we might discover naturally occurring a trend which could be defined mathematically, say in loss of readership of a given magazine, or loss of an audience for a television show, or loss of support for a political candidate. However, these movements are usually non-systematic, and thus fall into some other class.

Class 3

Here we have the case where the latent probability does not change but the latent position does, in some unsystematic way. The photograph is made under identical conditions, but the object being photographed changes in some way which cannot be described systematically by a mathematical equation. This situation is quite common in non-experimental research.

Classes 4, 5, and 6

As noted above, inductive regularities in the change of latent probabilities over time have frequently been noted, and many examples will be given below. These regularities are related to theory in various ways, but I have not been able to define them with mathematical precision. In some of the examples, no change is required in the latent class position of any respondent; others show trends which cannot be described on the basis of the data at hand in any precise way, although they may be approximately linear in some cases; and others are totally without any apparent systematicity, apparently being influenced by exogeneous factors acting on the variable.

For class 4 of these models, the photographic analogy would be, for a situation in which the latent probabilities are increasing systematically over time, a series of photographs in which the object photographed does not change over time, but the development of the film increases steadily in resolution and contrast, so that the photograph shows the object with increasing clarity. A low latent probability would correspond to a fuzzy picture, while a high latent probability would correspond to a picture of high resolution. Now, even though I have generally not been able to produce mathematical models which describe with precision the degree of latent probability, the regularities found frequently in these situations, and particularly in class 6, where there is also regularity in the movement from one latent position to another, even though that may not be described mathematically, are related to theory and provide a large territory which is promising for future work.

This group of models, class 6, is like a situation in which the degree of resolution of the photograph is increasing (or decreasing) and in which the object being photographed is changing in an unsystematic way (although not perhaps without qualitative regularities which are meaningful). Class 5 would be similar except that the changes in the object can be described by some mathematical process.

Classes 7, 8 and 9

The word "unsystematic" in the present context, as has been noted, means only that a mathematical model has not been found which describes with some precision or exactness (although perhaps probabilistically) the situation. It does not connote a lack of regularity. Few quantitatively inclined sociologists, from one of the greatest, Durkheim, to modern population forecasters, and for that matter, even few economic forecasters, with very sophisticated models, have been successful in making numerically accurate forecasts of economic variables consistently, on a precise quantitative basis; yet their qualitative forecasts have often been invaluable in the formation of policy.

In the examples we have had available, classes 7 and 9 have proved to be of considerable value. Classes 7, 8, and 9 would correspond to a situation in which a photograph was taken and developed with varying degrees of resolution. With low latent probability, we would see "as through a glass, darkly", while with high latent probability we would see the object with some precision. The positions of the objects would appear to change in class 7, even though in fact they did not. If the latent positions did actually change, as in classes 8 and 9, we would have compounded a change in resolution with an actual movement of the object.

In fact, that is what usually happens. We have the situation which McNemar refers to, which I dealt with in my publication of 1951 and my dissertation (1955a) and which Coleman describes cogently in the first chapter of his 1964 volume *Models of Change and Response Uncertainty.* The problem is to develop models which permit the separation of the apparent change caused by the probabilistic nature of the relation between the latent and the manifest response, on the one hand, and the change from one latent position to another, on the other hand.

The general situation is that the change appearing as a result of the probabilistic relation between the latent and manifest usually swamps the change from one latent position to another in the fields of non-experimental research (and even, to a lesser degree, in the quasi-experimental field research) we are considering, such as attitude-opinion research, media research, market research, educational research, economic anticipation research, and the like. This is the reason that in most examples the time interval makes very little or no difference. It is also one important reason why the stydy of the movement of a single variable over time can be very useful. Nevertheless, there is generally some movement from one latent position to another, and this will vary with the structure of the situation.

Coleman (1964) made the distinction between "learning" and "non-learning" models in this area, and also gave a good resume of some of the models which had been offered in this field, including the Blumen *et al.* (1955) "mover-stayer" model, the use of higher than first-order Markov chain processes, the Bush—Mosteller learning models, and the kind of models I had propounded. He neglected to note that I had dealt at some length (Wiggins, 1951, 1955a) with continuous space models, as will be seen also in later chapters of this monograph (limited, unfortunately, mainly to the normal distribution). He, on the other hand, developed a series of models based on the notion of continuous-time processes, which are quite different in a number of ways. His models also deal with problems arising from the effect of one variable on another.

However, he notes in a footnote on p. 11 of his 1964 monograph that under certain conditions a discrete-time Markov process may be viewed as embedded in a continuous-time process. The reader is referred to Barlow *et al.* (1965) for a more extended account of this relationship. They call this whole larger class of processes "semi-Markovian".

In the areas with which we are dealing here, except possibly for a few indicators such as the Neilson devices on television sets, the data come in discrete-time form. Thus although time itself is continuous, and on the face of it would seem to lend itself primarily to ordinary differential calculus and differential equations, the data lend themselves most conveniently in most situations to algebraic approaches.

With very rare exceptions, such as the one I located for a portion of the Lazarsfeld *et al.* study of voting intentions in Erie County, Ohio, (Lazarsfeld *et al.*, 1948), which was cited by Anderson (1951) and by almost everyone since who has written in the large literature on the use of Markov chains in social research, it turns out that the Markov chain does not hold on the manifest level. The fundamental assumption (one of two) of the Markov chain is that the future is wholly contained in the present; the past is irrelevant (of course, the past can be encapsulated into the present by higher-order processes, but as Coleman and I have noted this is a rather futile pursuit.) In general, I had shown (an example is given below) that on the manifest level one can never escape history. Kuehn (1958) later obtained a similar finding in his studies of brand loyalty. Of course, this whole question loomed very large in the philosophy of history, which is beyond our purview here.

The failure of the manifest Markov chain to describe empirical phenomena in field data was one reason for attempting other approaches. After all, two of the purposes of science were always described as "prediction and control". I myself have developed a number of other models for the description of panel data (Wiggins, 1954a,b,c). [1] These do not necessarily involve mathematics other

---

[1] There are a number of other statistical and mathematical approaches to the study of panel data, some of which are mentioned in this essay. One of my models for continuous space, for example, (1954b), involves the notion of each successive time point as including a component of an unchanging manifest position, a forgetting (extinction), and a new element, which may, if one wishes, be called the result of intervening experience. Most of the work in the field concerns itself primarily with the relationship of two or more variables over time, as in path analysis, much of the work of Coleman, a mode of analysis proposed by McFarland (1968) (which really covers both the single and the multiple variable situations), work by Boudon

than statistics. However, in many of the models of this monograph, use is made of the assumption that at some point the past becomes irrelevant on the latent level, although not on the manifest.

Among the models submitted here are those which provide for dependence on the latent level solely on the preceding time, on two preceding times, on three preceding times, and in principle so on ad infinitum, so we are not really bound to the Markovian assumption on the latent level. However, while the assumption that the entire past of a variable is totally embraced by a certain limited number of positions in past time may be useful on the latent level, it is curious that this encapsulation does not hold up on the manifest level. The swamping of most data by the probabilistic manifest—latent relation would, by the axiom of independence of the probability from one time to the next (similar to the assumption of local independence in latent structure analysis), seem to reduce the dependence on the past to zero even sooner on the manifest level.

There may be any one of a number of reasons for this situation. One is the possibility that the groups which are assumed to be homogeneous with regard to probability may not be completely homogeneous, as Coleman (1964) suggests. If even a handful of respondents have a response set, say, which leads to their invariably giving the same response, then their probability is not exactly the same as that of the others supposedly in the same group. In that case, the model would have to be complicated to add more groups. Of course, if everyone were exactly the same, there would be no social statistics other than number counts and no need for it. On the other hand, if everyone were entirely different, there would be no possibility of social science. Thus again the practical problem is to find reasonably parsimonious compromises which fit data.

One or two other considerations may be mentioned here. As Donald Campbell has noted in many contexts, regression on the mean does not depend on measurement error, although that may be a part of it, but in a situation standardized as to mean and variance and with a correlation over time less than perfect, it is a

(1968), and by Blalock, Duncan and others. I have made no attempt to cover this literature in the present work. In addition, as noted above, there are the many efforts to use the panel primarily to cumulate data, verify data, clean up data, and the like, as well as to study changes solely on the aggregative level. Also James Davis has a useful book on Panel Analysis.

statement of a tautology. From this he has shown, as have others in earlier but less complete analyses, that many of the seemingly substantive findings of some social scientists are artifacts.

The present models do not depend on the concept of measurement error, although they include it, but in portions of this essay the question of why the correlation is less than perfect is dealt with. Regression on the mean in general is, of course, not a mere tautology, since the mean and variance are not necessarily standardized by nature. The phenomenon of polarization is one of the most common social processes. Often "the rich get richer and the poor get poorer." A study reported by Bernard Weintraub in *The New York Times* in June, 1972, was done in England recently. It was based on a universe of 17,000 children born in Britain in one week in 1958. The study co-author, Ronald Davie, reported that "although children are tending to get taller each decade, the gap between the classes is not narrowing." The real problem is to create models to deal with cases of regression or non-regression, as the case may be. (I made a limited effort in this direction in Levenson and Wiggins, 1954).

The whole question of skewness of a variable is important in this connection, and I have dealt with that matter to some extent in Wiggins (1958) and in Chapter 7 below.

Finally, referring again to the scientific problem of "prediction and control", we should keep firmly in mind the all-important difference between causation and prediction. Some people think that no one has ever improved on Aristotle's classification of types of causes — formal, material, final, and efficient — and while Hume may have wakened Kant from his dogmatic slumbers, there are many who agree with Bertrand Russell that despite Mill and the hundreds of others who have debated the matter no one has ever refuted Hume's destruction of the concept of efficient cause. Whitehead said that modern science began when men stopped asking "why?" and started asking "how?".

Many recent workers in the field of mathematical models in the social sciences have devoted a good deal of time to questions of causation, and have produced valuable and interesting results. Ernest Nagel has shown in *The Structure of Science* (1961) the total confusion which has prevailed among many of the best minds on that subject in the field of the philosophy of science. The best prediction of an event or position in social science is almost invariably the position on the same variable at the smallest possible time interval prior to the predicted time. The best predictor of a man's vote is what he plans to do (ignoring measurement problems, secrecy, etc.) when he goes into the polling place. The best predictor of a student's grade average next year is his grade average this year. The best predictor of a job performance tomorrow is job performance today. The best predictor of whether a man will buy an electric razor tomorrow is whether he plans today to buy one tomorrow. The best predictor of a man's height tomorrow is his height today. Yet no one would call these predictors "causes" of the phenomena they predict.

To the regret of many, we have been able to predict and even control a great many things which we do not understand fully. In the long run, the more we understand, the more we can predict and control. Non-economists are now told that the best aggregate predictions over the long haul have been made by monetarists, who in general are reported to have the least theoretical understanding of any of the leading schools of economic thought why the relations they have found hold. In this work we have tried to avoid issues of this sort, which are much larger than those with which we have dealt.

Again, while I have nothing against simulations — indeed I have written (Wiggins, 1966) and lectured on the subject — it must be confessed that the extent to which they have been successful is profoundly limited in many social situations such as population projections, economic projections, etc., by the fact that the fundamental social situation, the attitudinal and behavioral accompaniments (such as the propensity to consume, the "women's liberation" movement) of the behavior which they were trying to predict, let alone the larger social and institutional movements such as wars, technology, court decisions, and so on, have overridden the rather simple assumptions and numbers on which they have had to be based. This by no means denigrates the value and potentiality of such work.

Of course, any model in this monograph can be simulated and any can be solved by computer methods. I have not entered into these matters but as noted in the Preface, have attempted in a limited area to present a set of models which cut across four areas — mathematical, methodological, several empirical fields of current use and interest, and theoretical. I believe this opens up many neglected aspects of longitudinal studies in the behavioral sciences, and enough models have been explicitly solved here and enough data shown to make me sanguine about the possibilities.

Certainly, no claim is made here for anything resembling the law of falling bodies. As Coleman (1964) justly observes, commenting on the type of problem with which the present work deals, "it is unlikely that any parsimonious model will be .... wholly satisfactory." Working on this problem for some twenty-five years, both mathematically and empirically, as I have done, does not necessarily provide the answers, or the correct answers, but it does guarantee awareness of the extreme complexity of the subject.

CHAPTER 2

Latent Probability Models with Fixed Probabilities and no Latent Change

The probabilities involved in a Markov chain are probabilities of moving from a given position at one time to another given position at a later time, and are thus appropriately called transition probabilities. The probabilities which form the basis for the models of this chapter are of a different sort. They do not directly refer to time or change, but rather to the relationship at a single point in time between a class to which in some sense a person, attitude, or item of behavior really belongs and a class into which it is placed by the data. It is the relationship between a latent class and a manifest class, although the terminology we shall employ is somewhat different.

Now the fact that the conceptualization leading to the models described in the present paper was arrived at independently and by a different route (namely, considerations of the reliability of qualitative data) by no means vitiates the fundamental linkage of this work to latent structure analysis. The notion of probabilities associated with latent and manifest classes has been so thoroughly expounded by Lazarsfeld and his associates (see, e.g., Lazarsfeld and Henry, 1968) that it needs little further explication here. It stems from the well-attested fact that any particular item of behavior is an imperfect indicator of more basic traits or attitudes or positions, partly because people are not perfectly consistent. A great many men who prefer blondes marry brunettes. The color of the wife's hair would be an imperfect indicator of the man's preference.

It is this imperfection of behavior items, their instability and complexity as concrete occurrences, which impels the search for more basic categorizations which are generally more stable, more abstract, and thus more serviceable for theory and prediction. The distinction between latent and manifest, between genotype and phenotype, has grown in importance. Latent structure analysis develops this distinction in terms of itemized tests. The responses to a series of associated items display a complex pattern of relationships which may be called the manifest structure of the data. This complex pattern is often derivable from a much simpler pattern of unobserved elements and relationships, which may be called the latent structure. The connecting link between latent classes and manifest data (responses to items) is a set of probabilities: a person falling into a given latent class has a certain probability of giving a particular response to a particular item. The basic assumption of latent structure analysis is called "local independence" and may be stated as follows: for persons within any latent class the items are independent (and therefore unrelated to each other), and hence the multiplicative law of probability holds. This law serves as the mathematical basis for solutions. For example, if within a given latent class the probability of giving a positive response to item i is a_i and the probability of giving a positive response to item j is a_j, then under the assumption the probability of giving positive responses to both items is $a_i a_j$. If all the classes are pooled, on the other hand, since the items are positively associated, the probability of a positive response on both items is greater than the product of the probabilities of the positive response on each item taken separately.

The present models are not extensions of latent structure analysis, but developments in a different direction using the concepts of latent probabilities and local independence. We are considering time series of panel data which refer only to a single item or variable. Latent structure analysis deals with many items taken at one point in time. We are dealing with one item taken at many points in time, although we shall later briefly comment on the two item panel. There is nothing to prevent the consideration of many items over many points in time, and a few beginnings have been made by others in that direction. Lazarsfeld (1950) considered the turnover in a latent dichotomy between two time points. Anderson (1951) commented on the possibility of a stationary Markov process applying to a latent structure rather than to a manifest item. Selvin and Simmel (1953) explored what happens to a fairly complex latent structure as the result of a propaganda campaign. But to push these efforts further is outside the scope of this paper.

Instead of talking of the latent structure of a set of items, we shall use the concept of the "latent structure" of a single item, or sometimes the "real" or "underlying" structure. The word "true" might be used here, for want of a better word, because there is a rough analogy here to what is called a "true score" in test psychology, as will be seen. It is not intended to convey any metaphysical implications.

The notion of latent structure of a single item may perhaps be best explicated by means of an example. Suppose in a panel studying a political campaign we ask respondents "Is your interest in the campaign great?" Respondents would answer "yes" or "no", but it is quite possible that neither of these two response groups is at all homogeneous in terms of interest in the campaign. There might be two relatively homogeneous groups among those who say "yes" and three relatively homogeneous groups among those who say "no". By "homogeneous group" we mean here intuitively people who "really" have the same degree of interest in the campaign.

This example is perhaps adequate as a first approximation, but more realistically we know that differences in the "real" positions will be reflected in the responses. Among respondents of moderate interest, for instance, some will respond "yes" and some "no" to the question about great interest. Even among those whose interest is basically or usually high, some will say "no" for any one of a number of reasons. Thus the relation between the latent position and the manifest response is not merely that any particular manifest response may correspond to more than one latent position, but also that any latent position may give rise to more than one manifest response. In what sense, then, is the latent class (those occupying a given latent position) homogeneous when some give one manifest response and some another? The latent class is homogeneous in the sense that each person in the class has the same *probability* of giving a particular manifest response. Probabilities of this kind are called latent probabilities. If we specify the latent classes, the proportion of respondents in each latent class, and the latent probabilities associated with each latent class, we have specified in full the "latent structure" of a single item.

At this point we need to consider briefly the relation of the concept of latent probability to the traditional concept of reliability. This is not the place to enter into an extended discussion of probability; the reader is referred to Nagel's short monograph for an excellent introduction to the frequency theory (Nagel, 1939), which Lazarsfeld adopts (Lazarsfeld, 1954), and to Cramer's *Mathematical Theory of Statistics* (1946) for the views of the school which introduces the probability of an event simply as a number associated with that event, but does not postulate the existence of concrete limits of frequency ratios. It is pertinent, however, to cite the example Lazarsfeld gives to illustrate the notion of probability, since this is similar to the definition of reliability of qualitative data which is given by most writers.

Lazarsfeld, in "A Conceptual Introduction to Latent Structure Analysis" (1954), says the following:

> Now it so happens that it is possible to assign a probability to Tom Brown individually.... Suppose we ask him several times the question, "Do you expect a third world war?" Suppose that each time the question is asked we "wash his brain", so that he forgets what answer he gave at previous interviews. Common sense (and actual experiments) lead us to expect that Tom Brown will sometimes say "Yes" and sometimes "No". We can then compute the proportion of times he says "Yes", as the number of interviews increases and thus establish the probability of a positive answer, along the lines of reasoning just developed. Such a probability would be a useful measure of whether Tom Brown is pessimistic regarding the political future. Here the Reference class is the set of repeated interviews under "brain-wash" conditions. This is, of course, an idealization which in concrete situations can only be approximated.

Let us first clarify one point, by noting that there are two factors which might make such a probability spurious — memory and change. Now the "brain wash" eliminates the factor of memory, but we need to assume also that the questions are repeated over a very short period of time so that no real change in Brown's attitude can occur during the interval. With this further condition, which Lazarsfeld really implies, his definition of probability becomes the basis of what is ordinarily called reliability. Reliability refers to the consistency of responses under conditions where neither memory nor real changes supervene. The formula which would ordinarily be used for the example cited is

$$p_{ii} = 2|p_i - 0.50|$$

where p_{ii} = reliability of item, p_i = probability of positive response

and the vertical marks on the right side mean the absolute value. If the probability of Brown's saying "Yes" is 1.00, he is perfectly consistent and the reliability is 1.00. If his probability is 0.50, he is perfectly inconsistent and the reliability is zero. The probability itself is essentially what is usually called "percentage agreement."

The more sophisticated models which we shall develop take care in good part of the factors of memory and change, as we shall see. Yet the latent probabilities are not reliability coefficients, in the ordinary sense of the word, for two reasons. The first reason has to do with the relation between the latent classes and the manifest classes. Suppose that for the item given by Lazarsfeld as an example there are three true classes, one in which there is a probability of 0.90 of saying "Yes", one in which there is a probability of 0.10 of saying "Yes", and one in which there is a probability of 0.50 of saying "Yes". The last is the one into which Tom Brown falls. Now it makes sense to say that the first class is the true "Yes" class and has a reliability of 0.90, and to say that the second class is the true "No" class and has a reliability of 0.90, but it does not make sense to say that the third class has a reliability of 0.50, because it does not correspond to any manifest class. In other words, the notion of reliability only makes obvious sense when the latent classes can be put into a one-to-one correspondence with the manifest classes. In many of the models herein this is done, and in the discussion we shall, in those cases, sometimes use the word "reliability" to refer to the latent probability that a person in a given latent class will give a manifest response in the corresponding manifest class. But the concept of reliability does not cover many of the situations which we shall consider, because there is often not a one-to-one correspondence between the latent classes and the manifest classes.

The second reason why the latent probabilities are not reliability coefficients, where the concept of reliability is considered to refer to error of measurement, is that much of the random change in items turns out to be behavioral rather than a consequence of the measure used. It often corresponds to what has been called "function fluctuation" in test psychology. In some of the examples which will be given, the data are of the sort which are known to be almost perfectly reliable, in the sense of error of measurement. Examples are the readership of magazines, or the grades of medical students. The data still reveal a strong random factor which behaves mathematically like an error of measurement but is demonstrably not such. The existence of this factor is important for the development of models of human behavior, even though it may be disappointing to those who wish to base a theory of behavior on the concept of rationality. Fortunately, this is not the aim of most sociologists or psychologists.

In addition to a random factor in behavior, there is usually an error of measurement, or more strictly, an error of classification, in the case of qualitative items. Mathematically, this kind of error is extremely difficult to distinguish from randomness in behavior or attitude.

In passing, it might be noted that the presence of a random factor in behavior or responses does not imply that individual behavior cannot be predicted. We are here working on the aggregative level, in situations where probability statements apply, but to say that a certain behavior is probable does not necessarily mean that it is unpredictable on another level. A factor which is random from one point of view or in one context may be predictable from another point of view or in another context. Thus even the definition of error of measurement is pragmatic and depends on the purpose of a given study (see Cronbach, 1947). What is error of measurement in one research might be the object of study or criterion in another. Coding errors might be highly predictable if the physiologist makes a detailed study, but they are random from the point of view of the sociologist who is interested only in the results of coding and not in the eye and hand movements, fatigue, etc. of the coders. Patricia Kendall (1954) has shown that many shifts in attitude are correlated with the mood of the respondent; but since the mood itself shifts randomly and is of no interest to the investigator in most situations, such attitude shifts are still usually defined as errors of measurement. For Kendall's study, however, they are the criteria. As Nagel (1939) says,

> It is a mistake to suppose that the successful use of probability statements depends in any way upon the issues of what is popularly known as "determinism" .... It is perhaps sufficient to note that the use of probability statements requires no commitment, even by implication, to any wholesale "deterministic" or "indeterministic" world-view; they can be used successfully in such contexts in which specified properties occur with stable relative frequencies in specified classes of elements.

It is now convenient to develop a notation for the latent structure. Let us designate the latent classes by the symbols C_1, C_2,..., C_{m}, and the proportions of the sample falling into each of these classes by v_1, v_2...., v_m. For any given class, C_i, let us denote the probability that members of this class will be reported in the positive response category by a_i. If we let p_1 represent the proportion of the total sample which is reported in the positive response category (e.g., proportion responding "Yes"), we have the relation

$$p_1 = a_1 v_1 + a_2 v_2 + \dots + a_m v_m \tag{2.1}$$

or in vector notation

$$p_1 = a'v \tag{2.2}$$

This relationship holds because we assume that each latent class is homogeneous in the sense that each person in the class has the same probability of giving the positive response, and we use the proportion positive in the sample as an estimate of the probability of giving a positive response (the equation does not take account of sampling variation, as noted below).

Let us call the changes from one latent class to another the "latent" changes. There are two ways in which we could represent these changes. We could set up a matrix of transition probabilities, as in the case of the Markov chain, which would show the probability that a person in a given true class C_i at time 1 changes to any given true class C_j at time 2. It is more convenient for our present purposes, however, to consider all those respondents in class C_i at time 1 (the initial time) and in class C_j at time 2 as constituting a latent class C_{ij}, and to designate the proportion of the total sample falling into the class C_{ij} by v_{ij}. Members of class C_{ij} then have the probability $a_i a_j$ of being reported in the positive response category at both time 1 and time 2. This is the principle of local independence coming in under another guise; we know that for the whole sample the responses at time 1 and time 2 are positively related, but we assume that within a given latent class they are unrelated.

If we let p_{11} be the proportion of the sample which according to the manifest data gives the positive response at both time 1 and time 2, we get the relation

$$\begin{aligned} p_{11} = {} & a_1 v_{11} a_1 + a_1 v_{12} a_2 + \dots + a_1 v_{1m} a_m \\ & + a_2 v_{21} a_1 + a_2 v_{22} a_2 + \dots + a_2 v_{2m} a_m \\ & + \dots\dots\dots\dots\dots\dots\dots\dots \\ & + a_m v_{m1} a_1 + a_m v_{m2} a_2 \quad + \dots + a_m v_{mm} a_m \end{aligned} \tag{2.3}$$

or in matrix notation

$$p_{11} = a'Va \tag{2.4}$$

We shall use equations of this type to obtain solutions, and it must be stated again that sampling variation is not taken into account in this paper. We assume that the data are "infallible", i.e., that the sample provides a perfectly accurate estimate of the universe. In the development of models, it often appears to be good strategy to worry about statistical considerations rather late in the game.

We have, then, the following basic notation:

Time	*Latent class*	*Proportion in latent class*	*Proportion giving manifest response*	*Latent probability*
1	$C_{i.}$	$v_{i.}$	$p_{1.}$	a_i
1, 2	C_{ij}	v_{ij}	p_{11}	$a_i a_j$

The manifest proportions are part of the data; the true classes, proportions in true classes, and latent probabilities are the item latent structure and are inferred.

We shall consider a number of models of this general sort. We shall start with the simplest of these models, and go on to the more complicated, which are probably more realistic. All the models presented in the text will be based on dichotomous manifest items, although this restriction is arbitrary. Solutions have been obtained for models based on items with any number of manifest categories, and the basis for these solutions is given in Appendix B.

We begin with the simplest models, in which it is assumed that no latent change and no change in latent probabilities occur. We go on to Chapter 3 to models in which it is assumed that latent change occurs in only one direction, and then to models where it is assumed that latent change occurs in both directions. We then consider, in Chapter 4, a simple model in which the true changes are assumed to follow a stationary Markov chain process. In all these models it is assumed that the latent probabilities themselves do not change. In Chapters 5 and 6 we consider models in which the latent probabilities are assumed to change, and a variety of mixed models, in which both latent changes and changes in latent probabilities may occur.

It was mentioned above that the two factors of memory and latent change are taken into account by the models we shall develop. Considering these in reverse order, we may first note that latent change poses a problem not merely for latent probability models but also for traditional considerations of the measurement of reliability, as McNemar (1946) has remarked in his paper "Opinion-attitude methodology". He says that

> Some have objected to the second scheme [determining reliability of a single item by correlating answers to the same item at two different times] because it is feared that bona fide changes may have taken place; thus an individual would not tend to give the same answer twice. This poses a dilemma: If an opinion or attitude is such a momentary thing that individual reliability or consistency is lacking, how can one expect the opinion in question to correlate with stable sociological or psychological characteristics of the individual? The research problems associated with unstable as opposed to relatively stable opinions and attitudes must be quite different, yet we fail to find this distinction in the literature.

We deal in this study with these very unstable attitudes and opinions, as well as behaviors, which contrast very sharply with stable characteristics and traits. The simplest models proposed do not permit any "latent" or "true" changes, but in Chapter 3 we develop models which permit true change and which discriminate this kind of change from random fluctuation. The confounding effect of latent or "true" changes is removed, in so far as these changes are systematic and mathematically distinguishable from random changes. In a sense, the whole of this work represents suggested solutions to the problem posed by McNemar.

The other confounding factor, memory, has the effect of spuriously increasing the consistency of responses from one time to another. The models advanced here do not eliminate this factor entirely, but the more complex models do eliminate it in good part. To the extent that memory is memory of manifest responses, the confounding effect is not altogether eliminated. But to the extent that true positions are affected by memory of true positions at previous times, the models take account of the factor by allowing the latent position at a given time to depend on the latent positions at a specified number of previous times. The models which are proposed (and for which solutions are offered) permit any length of dependency on past time, within the limits of the available date, except that they usually require that the latent position at the final time point be independent of the latent position at the initial time point when the latent positions at all intermediate time points are held constant. Thus the confounding effects of memory are largely eliminated in the more elaborate models.

Finally, there is some consideration in these chapters of panel data which are in the form of continuous variables, usually assumed to have normal distributions. In such data we cannot speak of the probability of giving a certain response, since strictly speaking the probability of a particular manifest score or value is zero, and the number of latent classes is infinite, while the proportion in any latent position is zero. With these kinds of data we shall deal in terms of correlation matrices, where the scores at time 1 are correlated with the scores at time 2, and with the scores at time 3, and so on; and where the scores at each time are correlated with the scores at every other time. The treatment will usually be rather superficial, but we shall indicate what kind of matrix would result from a relation between the variable and itself at various time points which is analogous to that postulated for the discrete variables. This will help the reader to grasp the situation for discrete variables.

Cases Where No Latent Change Occurs

There are many items included in panel studies such that the contents of the items themselves virtually or wholly preclude latent change. This is most often true of factual items. Certain of these items, such as age, sex, and education, wholly preclude true change over a short time interval, or restrict true change to very narrow limits. Other items, such as socio-economic status, are likely to change very little over a short interval. There are also attitude questions where no marginal shift is observed in the responses, and where, in the absence of sufficient information to use a more complex model, it seems justifiable to assume that no latent change has occurred in order to obain an estimate of latent probability. Such items, of course, must be approached with caution. We shall consider first only cases where the assumption that there is no change from one latent class to another seems warranted.

Two Interviews, Two Latent Classes

Let us again refer to the notation to which we shall adhere with slight modifications throughout this chapter. The proportions giving various responses in the manifest data will be denoted by the letter "p" with n subscripts, and n is the number of time points or interviews involved in the case under consideration. The position of a subscript will indicate the time to which it refers. The subscript "1" will refer to the positive response, the subscript "2" to the negative response, and the subscript "." to the sum of all responses. If, as in the present case, there are two interviews, the proportion of the sample giving the positive response at time 1 will be $p_{1.}$. The proportion giving the positive response at both times is p_{11}.

Similarly, the proportion of the sample in latent class C_1 at time 1 will be denoted by $v_{1.}$, the proportion in class C_1 at time 2 will be denoted by $v_{.1}$, and the proportion in class C_1 at both times by v_{11}. Since there may be more than two latent classes, we generalize this notation so that the proportion of the sample in class C_i at time 1 is $v_{i.}$, and the proportion of the sample in class C_i at time 1 and class C_j at time 2 is v_{ij}. There are certain minor variations in notation from one model to another, and these will be indicated in each case.

If we have only two interviews at our disposal, we must make very strong assumptions. We have already assumed that there is no latent change. As a consequence of this, the manifest marginals may be expected to remain unchanged (ignoring, as always here, sampling variations). The manifest data, in fact, will have the following form:

		time 2 +	time 2 —	
time 1	+	p_{11}	p_{12}	$p_{1.}$
	—	p_{21}		
		$p_{.1}=p_{1.}$		1

Since $p_{.1} = p_{1.}$, there are only two independent pieces of information available, $p_{1.}$ and p_{11}. We must make assumptions which reduce our unknowns to three, since we always have the additional equation

$$\sum_{ij} v_{ij} = 1 \tag{2.5}$$

Let us assume that there are only two latent classes at time 1, c_1 and c_2. Since there is no latent change, we have

$$v_{1.} = v_{11} = v_{.1} \tag{2.6}$$

We further assume that the probability of positive response for one latent class is the complement of the probability of positive response for the other. For simplicity of solution to equations in this section, let us define as the probability of a positive response for individuals in class C_1 the sum "$a + b$", while the letter "a" alone will mean the probability that an individual in class C_1 will give the negative response.

We then have by assumption of this model,

$$a + b = 1 - a \tag{2.7}$$

where $a + b$ is the probability of positive response for those in the positive true class which by assumption equals the probability of negative response for those in the negative true class. Schematically, we have the following structure:

Latent class	*Proportion in latent class*	*Probability of positive response, time 1*	*Probability of positive response, time 2*
C_1	$v_{1.}$	$a + b = 1 - a$	$a + b = 1 - a$
C_2	$v_{2.}$	a	a

By (2.1), we have

$$\begin{aligned} p_{1.} &= (a+b)v_{1.} + av_{2.} \\ &= a(v_{1.} + v_{2.}) + bv_{1.} \\ &= a + bv_{1.} \end{aligned} \tag{2.8}$$

Similarly, we get

$$p_{.1} = a + bv_{.1} = a + bv_{1.} \tag{2.9}$$

By (2.2), we have

$$\begin{aligned} p_{11} &= (a+b)^2 v_{11} + a^2 v_{22} \\ &= a^2(v_{11} + v_{22}) + 2abv_{11} + b^2 v_{11} \\ &= a^2 + 2abv_{1.} + b^2 v_{1.} \end{aligned} \tag{2.10}$$

Let us introduce the notation

$$[12] = p_{11} - p_{1.}p_{.1} \tag{2.11}$$

Multiplying (2.8) by (2.9), we get

$$p_{1.}p_{.1} = a^2 + 2abv_{1.} + b^2 v_{1.}^2$$

Hence

$$\begin{aligned} [12] &= b^2 v_{1.} - b^2 v_{1.}^2 \\ &= bv_{1.}[b(1 - v_{1.})] \end{aligned} \tag{2.12}$$

Let

$$y = bv_{1.}$$

and

$$x = b(1 - v_{1.})$$

Then

$$[12] = xy$$

By (2.7)

$$2a = 1 - b$$

By (2.8)

$$\begin{aligned} 1 - 2p_{1.} &= 1 - 2a - 2bv_{1.} \\ &= 1 - (1 - b) - 2bv_{1.} \\ &= b - 2bv_{1.} \\ &= b(1 - v_{1.}) - bv_{1.} \\ &= x - y \end{aligned}$$

Solving for x and y, we get

$$x = (\tfrac{1}{2} - p_{1.}) \pm \sqrt{(\tfrac{1}{2} - p_{1.})^2 + [12]}$$

$$y = -(\tfrac{1}{2} - p_{1.}) \pm \sqrt{(\tfrac{1}{2} - p_{1.})^2 + [12]}$$

Since we have arbitrarily designated the latent classes in such a way that class C_1 has a greater probability of positive response than C_2, we have

$$a + b > a$$

$$b > 0$$

Also,

$$v_1 > 0$$

and

$$1 - v_1 > 0$$

Hence

$$x = b(1 - v_1) > 0$$

$$y = bv_1 > 0$$

Since in general $[12] > 0$, we shall have

$$|\sqrt{(\tfrac{1}{2}-p_{1.})^2 + [12]}| > |(\tfrac{1}{2}-p_{1.})|$$

Therefore, in order to make x and y positive, we must use the positive root of the expression under radicals in both x and y. From (2.8) we have

$$\begin{aligned} a &= p_{1.} - bv_{1.} \\ &= p_{1.} - y \end{aligned}$$

$$a = \tfrac{1}{2} - \sqrt{\tfrac{1}{4} + p_{11} - p_{1.}} \qquad (2.13)$$

From (2.7) and (2.13)

$$a + b = 1 - a = 1 - [\tfrac{1}{2} - \sqrt{\tfrac{1}{4} + p_{11} - p_{1.}}]$$

$$a + b = \tfrac{1}{2} + \sqrt{\tfrac{1}{4} + p_{11} - p_{1.}} \qquad (2.14)$$

Finally, from (2.8), we have

$$v_{1.} = \frac{p_{1.} - a}{b} = \frac{p_{1.} - a}{1 - 2a} \qquad (2.15)$$

Example of Two Interviews, Two Latent Classes Assumed

Few panel data are available for which the assumption that there are only two latent classes seems entirely reasonable. In order to illustrate the method, we shall apply it to an example where the assumption will be made nonetheless. The example is taken from the 1948 voting study done in Elmira, New York (see Berelson *et al.*, 1954); the item refers to the education of respondents. The two interviews were conducted in June and October, 1948. If we dichotomize the sample into respondents who did not graduate from high school (not HSG) and those who did (HSG), we have the data illustrated in Table 1.

Observe that there is a net decrease of 16 in the number of those who are reported as high school graduates. Since it is impossible for a person's education to decrease, this decrement is probably caused by sampling variation. The sampling distribution here refers to all possible manifest distributions which would be produced by variable error affecting the distribution of the respondents in the panel. The manifest distribution at any one time is a

TABLE 1

		October		
		Not HSG	HSG	Total
June	Not HSG	358	53	411
	HSG	69	317	386
	Total	427	370	797

sample from this distribution. From one interview to another there will be sampling variation, even if the latent distribution remains unchanged. The sample of respondents, of course, does not change. In such cases the best procedure is to take for $p_{1.}$ the mean of $p_{1.}$ and $p_{.1}$. We get

$$p_{1.} = \frac{411 + 427}{2(797)} = 0.526$$

$$p_{11} = \frac{358}{797} = 0.449$$

$$\begin{aligned} a &= \tfrac{1}{2} - \sqrt{\tfrac{1}{4} + p_{11} - p_{1.}} \\ &= 0.50 - \sqrt{0.173} \\ &= 0.50 - 0.416 \\ &= 0.084 \end{aligned}$$

By (2.7),

$$a + b + 1 - a = 0.916$$

And by (2.15), we have

$$v_{1.} = \frac{p_{1.} - a}{1 - 2a} = \frac{0.526 - 0.084}{1 - 0.168} = 0.531$$

Notice that the latent proportion $v_{1.}$ is slightly larger than the manifest proportion $p_{1.}$. We shall want to comment on this later.

We find in this example that there is a probability of 0.916 that a person will be correctly classified as a high school graduate or not (assuming always the absence of systematic or biasing error). Although we shall adhere to the use of probabilities as measures of reliability as well as latency, it is perhaps instructive to keep in mind the approximate relation that exists in the present case between these probabilities and some more commonly used measure of reliability, such as the point correlation (remembering that by dichotomizing the responses we have reduced the correlation). In (2.12) we had

$$[12] = b^2 v_{1.}(1-v_{1.}) ,$$

which gives

$$b^2 = \frac{p_{11}-p_{1.}p_{.1}}{v_{1.}(1-v_{1.})} = \frac{p_{11}-p_{1.}p_{.1}}{v_{1.}v_{2.}}$$

Since in general $v_{1.}$ will not differ much from $p_{1.}$, $v_{2.}$ will not differ much from $p_{2.}$. Since there is no latent change, and very little change in the manifest marginals, we have the following set of approximations:

$$v_1 \cdot v_{2.} = \sqrt{v_{1.}v_{2.}v_{.1}v_{.2}} \simeq \sqrt{p_{1.}p_{2.}p_{.1}p_{.2}}$$

Hence we have the following rough approximation:

$$b^2 = (1-2a)^2 \simeq \frac{p_{11}-p_{1.}p_{.1}}{\sqrt{p_{1.}p_{2.}p_{.1}p_{.2}}}$$

The expression on the right, however, is nothing but the point correlation between the responses at time 1 and time 2. In order to get an approximation of this correlation, it is merely necessary to square b. For the example we have given, the actual point correlation is 0.692, while b^2 is also 0.692. Of course, the approximation will not always be this close.

Three Interviews, Two Latent Classes

We next consider a case where three interviews are taken, and where it is reasonable to assume that no latent change has occurred and that at any given time there are only two latent classes. Since there is no latent change, we have the following relations:

$$v_{1..} = v_{.1.} = v_{..1} = v_{11.} = v_{1.1} = v_{.11} = v_{111} \tag{2.16}$$

For convenience, let us call this proportion v.

We have also the relations

$$p_{1..} = p_{.1.} = p_{..1} \tag{2.17}$$

and

$$p_{11.} = p_{1.1} = p_{.11} \tag{2.18}$$

In the present case, we have one more item of information than we had in the previous case, namely, p_{111}, hence we do not need to make any assumption about complementarity. We have the following structure:

Latent class	*Proportion in latent class*	*Probability of positive response*		
		Time 1	*Time 2*	*Time 3*
C_1	v	$a+b$	$a+b$	$a+b$
C_2	$1-v$	a	a	a

We have the following relations:

$$\begin{aligned} p_{1..} = p_{.1.} = p_{..1} &= (a+b)v + a(1-v) \\ &= a + bv \end{aligned} \tag{2.19}$$

Letting $[12] = p_{11.} - p_{1..}p_{.1.}$, etc., we have, by (2.12), (2.17) and (2.18),

$$[12] = [23] = [13] = b^2v(1-v) \tag{2.20}$$

Letting

$$y = bv$$

and

$$x = b(1-v)$$

as before, this becomes

$$[12] = [23] = [13] = xy \tag{2.21}$$

We have in addition,

$$\begin{aligned} p_{111} &= (a+b)^3v_{111} + a^3v_{222} \\ &= a^3(v+1-v) + 3a^2bv + 3a^2v + b^3v \\ &= a^3 + 3a^2bv + 3ab^2v + b^3v \end{aligned} \tag{2.22}$$

We shall now use an expression [123], which we define as follows:

$$\begin{aligned} [123] &= p_{111} - p_{1..}[23] - p_{.1.}[13] - p_{..1}[12] - p_{1..}p_{.1.}p_{..1} \\ &= p_{111} - 3p_{1..}[23] - p_{1..}{}^3 \end{aligned} \tag{2.23}$$

If we make use of (2.22), (2.20), and (2.19), this expression becomes

$$[123] = b^3v(1-v)^2 - b^3v^2(1-v) \qquad (2.24)$$

$$= x^2y - xy^2$$

Since $[12] = xy$, we divide (2.23) by [12], getting

$$\frac{[123]}{[12]} = x - y \qquad (2.25)$$

Solving (2.21) and (2.25) for x and y gives

$$x = \frac{[123]}{2[12]} \pm \sqrt{\left(\frac{[123]}{2[12]}\right)^2 + [12]}$$

$$y = \frac{-[123]}{2[12]} \pm \sqrt{\left(\frac{[123]}{2[12]}\right)^2 + [12]}$$

Since we have arbitrarily designated the latent classes in such a way that class C_1 has a greater probability of positive response than C_2, we have

$$a + b > a$$

$$b > 0$$

Also,

$$v > 0$$

and

$$1 - v > 0$$

Hence

$$x = b(1-v) > 0$$

$$y = bv > 0$$

Since in general $[12] > 0$, we shall have

$$\left|\sqrt{\left(\frac{[123]}{2[12]}\right)^2 + [12]}\right| > \left|\frac{[123]}{2[12]}\right|$$

Therefore, in order to make x and y positive, we must use the positive root of the expression under radicals in both x and y.

By (2.19)

$$a = p_{1..} - y$$

$$a = p_{1..} + \frac{[123]}{2[12]} - \sqrt{\left(\frac{[123]}{2[12]}\right)^2 + [12]} \qquad (2.26)$$

Also from (2.19)

$$a + b = p_{1..} + x$$

$$a + b = p_{1..} + \frac{[123]}{2[12]} + \sqrt{\left(\frac{[123]}{2[12]}\right)^2 + [12]} \qquad (2.27)$$

Finally, from (2.19)

$$v = \frac{p_{1..} - a}{b}$$

Example of Three Interviews, Two Latent Classes Assumed

The paucity of panel data collected for three waves often leads us to use examples in which the assumption that there is no latent change is probably unwarranted. These examples will be included here for their heuristic value, and also because two other models will later be postulated of the same data, for comparative purposes. We can then gauge the effect of using the present inadequate model.

The first set of data refers to the readership of various magazines, from which two have been selected which we shall call magazines "A" and "B". The raw data for magazine "A" are as follows:

TABLE 2

Readership of magazine "A"

time 1	reader		non-reader	
time 2	reader	non-reader	reader	non-reader
time 3: reader	30	11	17	64
time 3: non-reader	14	60	56	734

The pattern for magazine "B" is very similar. The proportions giving the positive response, i.e., reading the magazines, are as follows:

TABLE 3

	Magazine "A"	*Magazine "B"*
$p_{1..}$	0.124	0.121
$p_{.1.}$	0.119	0.116
$p_{..1}$	0.117	0.107
$p_{11.}$	0.048	0.047
$p_{1.1}$	0.042	0.042
$p_{.11}$	0.045	0.049
p_{111}	0.030	0.026

Since for each magazine the first order marginals are approximately equal at all three times, as are also the second order relations, the assumption of no latent change seems to be reasonably justified in the absence of other evidence. The simplest procedure for applying the formulas of the previous section to these data is to average the first order marginals for all three times, and to average the point probabilities for all three pairs of times, for each magazine. When this is done, and the equations are solved, we get the following results:

TABLE 4

	Latent class	*Proportion in latent class*	*Probability of positive response*
Magazine "A"	C_1	0.057	0.745
	C_2	0.943	0.082
Magazine "B"	C_1	0.099	0.605
	C_2	0.901	0.048

Inspection of the data apparently reveals that magazine "B" has more latent readers than magazine "A", 0.099 as against 0.057, while the latent probability of the item for the class of latent readers is greater for magazine "A" than for magazine "B", 0.745 against 0.605. It happens that we know from the way in which this study was conducted that the reliabilities of these questions are extremely high. Respondents were required to identify a certain proportion of the contents of the current issue of each magazine before being classified as readers, the interviewers were highly experienced, and the coding procedures were carefully checked.

The question of using the word "reliability" as a synonym for "latent probability" must be raised. What we actually have is undoubtedly a case of random variability in the behavior of reading the magazine, rather than unreliability in determining the readership. Our computations have not given us the probability that a person who has actually read a given issue of the magazine will be correctly classified as a reader, but a probability which is the product of that probability and the probability that the person who in some sense is a regular reader of the magazine will read any particular issue.

When we speak of a "regular" reader, there are two relevant points to be noted. First, the word has reference to something that occurs through time. Regularity of readership would have no meaning if applied to a single event. Yet the word has meaning as applied to a person at a given time. If we say that a person is a regular reader of a magazine, we do not necessarily mean that he has always been a regular reader, nor that he will be a regular reader forever. We mean that at the present time he is a person whose pattern of dispositions is such that it includes a high probability that he will read any given issue of the magazine. What is behavior, occurring through time, rests on the basis of an attitude, which may or may not remain stable throughout that time. If the attitude varied as randomly as the behavior, it would make no sense to talk of a "regular" reader, in the present context. If there is no stable factor in the respondent, no meaning can be given to the statement that at a given time "he has a certain probability of reading the magazine". In this sense his behavior through time can often be collapsed into a moment. Respondents may then be said to fall into various classes at a given time, classes which are homogeneous in the sense that each person in a class has the same probability of reading not merely a single issue but any one of several issues of the magazine. Such a probability, while it applies to the respondent at any particular time, refers to his behavior over a period of time. Any account of the readership of respondents which purports to be adequate must classify them, not according to their behavior at a single time, but according to the dispositions which cause their habitual behavior to differ. It is such differences between (inferred) dispositions which form the basis of the latent classes which we determine by the procedures advanced in this paper.

In the light of this, the question of using the word "reliability" as applied to the probabilities we determine becomes clear. If the term is meant in the ordinary sense as the probability that a person who is classified in some more or less arbitrary manifest class will be classified in the same way under identical conditions, then the word is misused if applied to latent probabilities.

It is clear that the latent probabilities of different individuals or even classes may themselves change over time. Both experience and data confirm this, and later chapters will discuss these changes and offer some rationale for their occurrences and the patterns of their occurrences which are often found.

The second point to be mentioned, which follows from the first, is that the assumption in the examples given that there are only two latent classes is somewhat dubious. While at any given time there are readers and non-readers, it is prima facie plausible that the habits and attitudes of the respondents divide them into several classes. There are probably some respondents who always read a given periodical, some who never read it, others who read it frequently, others who read it seldom, and so on. If we had many interviews, we could determine how many such latent classes there are. In the absence of further information, we can only apply several models and conjecture. We shall see later how the results from two other models compare with those obtained here.

Three Interviews, Three Latent Classes Assumed

There may be cases where three interviews are available with no apparent latent change, when for some reason the assumption that there are only two true classes seems unreasonable. A solution then may be obtained for three true classes if severe restrictions are placed on the latent probabilities. One model of this type may be briefly included here. This model is based on the assumption that the latent probability of class C_1 is the same as that of class C_3, and that the middle class C_2 is totally "neutral" with respect to the two possible responses. This gives the following schema:

Latent class	*Proportion in latent class*	*Probability of positive response*		
		Time 1	*Time 2*	*Time 3*
C_1	v_1	$0.5 + a$	$0.5 + a$	$0.5 + a$
C_2	v_2	0.5	0.5	0.5
C_3	v_3	$0.5 - a$	$0.5 - a$	$0.5 - a$

We have the relations

$$p_{1..} = (0.5+a)v_1 + 0.5v_2 + (0.5-a)v_3$$

$$= 0.5 + a(v_1 - v_3) \tag{2.28}$$

$$a(v_1 - v_3) = p_{1..} - 0.5$$

$$p_{11.} = (0.5+a)^2 v_1 + 0.5^2 v^2 + (0.5-a)^2 v_3 \tag{2.29}$$

$$= 0.25 + a(v_1 - v_3) + a^2(v_1 + v_3)$$

By (2.28) and (2.29) we get

$$a^2(v_1 + v_3) = p_{11.} - p_{1..} + 0.25 \tag{2.30}$$

We have also

$$p_{111} = (0.5+a)^3 v_1 + 0.5^3 v_2 + (0.5-a)^3 v_3 \tag{2.31}$$

$$= 0.125 + 0.75a(v_1 - v_3) + 1.5a^2(v_1 + v_3) + a^3(v_1 - v_3)$$

Substituting (2.28) and (2.30) in (2.31)

$$a^2 = \frac{p_{111} - 1.5p_{11.} + 0.75p_{1..} - 0.125}{p_{1..} - 0.5} \tag{2.32}$$

The probabilities are easily determined from a. Also, from (2.28) and (2.30),

$$v_1 - v_3 = \frac{p_{1..} - 0.5}{a}$$

$$v_1 + v_3 = \frac{p_{11.} - p_{1..} + 0.25}{a^2}$$

The sum of these gives $2v_1$, and their difference gives $-2v_3$.

Example of Three Interviews, Three Latent Classes Assumed

Let us briefly consider the result of applying the present model to the magazine data used previously. The outcome is presented in Table 5.

With the third class included, the probabilities of positive response are much more similar for the two magazines than they were in the two class case. There they differed by 0.140 for class C_1 and 0.034 for class C_2; here they differ by 0.025 for both class C_1 and class C_3. Class C_2 has the probability of 0.5 by assumption

TABLE 5

	Latent class	Proportion in latent class	Probability of positive response
Magazine "A"	C_1	0.026	0.934
	C_2	0.072	0.500
	C_3	0.902	0.066
Magazine "B"	C_1	0.010	0.959
	C_2	0.140	0.500
	C_3	0.850	0.041

in both cases, of course. If we call the members of class C_1 regular readers, and the members of class C_2 occasional readers, we see that magazine "A" has about two and one half times as many regular readers as magazine "B", while magazine "B" has about twice as many occasional readers as magazine "A". Magazine "A" would thus be a much better medium for a continuing advertising campaign in which it was desired to have the same persons read a number of advertisements appearing in successive issues. In support of the results we may mention that magazine "A" carries novels in serial form, while magazine "B" does not, a fact which would make for more regular readership of magazine "A".

The Use of Three Interviews to Check Assumptions of the Two-Interview Model

It is desirable to check assumptions if possible, and one primary value of additional data is that when applied to models requiring less data for a determinate solution the applicability of the model may be tested. Either the model may be applied to a part of the data, and an attempt made to extrapolate the remaining data from the parameters of the model, or else the conditions of reducibility of the more complex data to the simpler model may be found. In either case, the statistical problem remains, since without tests of significance there are no precise means for determining how close a given fit should be in order to be considered "good". Since this paper does not often deal with statistical problems, the question must in most cases remain open. There still remains an heuristic value in discovering conditions of reducibility.

If three interviews are available, the applicability of the first model described may be tested. That model assumes that there are two latent classes, each of which has the same reliability. If it is now assumed that the same model holds over three interviews, the necessary condition for this to be true may be found.

Let

$0.5 + a$ = probability of positive response for class C_1
$0.5 - a$ = probability of positive response for class C_2
$t = v_1 - v_2$ = difference between proportions of sample in the two latent classes

Then

$$p_{1..} = p_{.1.} = p_{..1} = (0.5 + a)v_1 + (0.5 - a)v_2 \quad (2.33)$$
$$= 0.5(v_1 + v_2) + a(v_1 - v_2)$$
$$= 0.5 + at$$

Similarly,

$$p_{11.} = p_{1.1} = p_{.11} = 0.25(v_1 + v_2) + a(v_1 - v_2) + a^2(v_1 + v_2)$$
$$= 0.25 + at + a^2 \quad (2.34)$$
$$p_{111} = 0.125 + 0.75at + 1.5a^2 + a^3t \quad (2.35)$$

From (2.33) and (2.34),

$$p_{1..} - p_{11.} = 0.25 - a^2$$
$$a^2 = -p_{1..} + p_{11.} + 0.25 \quad (2.36)$$

Subtracting (2.33) multiplied by $(a^2 + 0.75)$ from (2.35),

$$p_{111} - a^2 p_{1..} - 0.75p_{1..} = -0.25 + a^2$$

or

$$a^2 = \frac{p_{111} - 0.75p_{1..} + 0.25}{1 + p_{1..}} \quad (2.37)$$

Equating (2.36) and (2.37),

$$-p_{1..} + p_{11.} + 0.25 = \frac{p_{111} - 0.75p_{1..} + 0.25}{1 + p_{1..}}$$

This reduces to

$$p_{1..} = \frac{p_{11.} - p_{111}}{p_{1..} - p_{11.}} \quad (2.38)$$

In determinantal form, it is

$$\begin{vmatrix} 1 & p_{11.} \\ p_{1..} & p_{111} \end{vmatrix} = \begin{vmatrix} 1 & p_{1..} \\ p_{1..} & p_{11.} \end{vmatrix} \qquad (2.39)$$

In practice, we may apply to three-interview data the more complex model which puts no restrictions on the latent probabilities, and observe whether they are the same or different for the two classes, or we may first note whether the equation (2.38) holds for the data. If it does not hold to some (here unspecified) degree of approximation, we cannot use the simpler model. If we examine the magazine data used previously, we find that for magazine "A" the left hand side of (2.38), for which it is preferable to use the mean of the manifest marginals rather than $p_{1..}$ alone, is 0.12, while the right hand side is 0.16, a difference of 0.04. For magazine "B" the left hand side is 0.11, and the right hand side is 0.29, a difference of 0.18. Thus the model fits the data of magazine "A" dubiously, while it seems to fit magazine "B" very badly. When we remember that the two probabilities obtained without restriction for the same data were much more similar for the two classes for magazine "A" than for "B", the result seems reasonable.

Two Models for Four Interviews

With four interviews we can determine three latent classes by putting one restriction on the probabilities. For example, we could assume that the probabilities of the extreme classes are complementary, with the middle class having an unknown probability. We could also solve for four latent classes if we put such restrictions on the probabilities as to limit the unknowns to one. One model of each type will be outlined. The three-class model has the following structure:

Latent class	*Proportion in latent class*	*Probability of positive response*			
		Time 1	*Time 2*	*Time 3*	*Time 4*
C_1	v_1	$0.5 + a$	$0.5 + a$	$0.5 + a$	$0.5 + a$
C_2	v_2	k	k	k	k
C_3	v_3	$0.5 - a$	$0.5 - a$	$0.5 - a$	$0.5 - a$

We have assumed here that there are three latent classes, and that the latent probability is related to the class in such a manner that extreme positions have complementary probabilities, while the middle position has an unspecified probability, which we expect to be between the other two. There is a great deal of evidence that this kind of assumption often is better warranted than such an assumption as linearity of the probabilities.

Let us make temporarily the following substitutions:

$$y = 0.5 + a$$

$$z = 0.5 - a$$

Then consider the product

$$P = \begin{pmatrix} v_1 & v_2 & v_3 \\ v_1 y & v_2 k & v_3 z \\ v_1 y^2 & v_2 k^2 & v_3 z^2 \end{pmatrix} \begin{pmatrix} 1 & y(1-y) & 0 \\ 1 & k(1-k) & 1/v_2 \\ 1 & z(1-z) & 0 \end{pmatrix} \tag{2.40}$$

If we multiply these matrices and substitute manifest data, we get

$$P = \begin{pmatrix} 1 & [p_1 - p_{11}] & 1 \\ p_1 & [p_{11} - p_{111}] & k \\ p_{11} & [p_{111} - p_{1111}] & k^2 \end{pmatrix} \tag{2.41}$$

This may be seen easily as follows. Consider for example the term $p_1 - p_{11}$. We have

$$\begin{aligned} p_1 - p_{11} &= v_1 y + v_2 k + v_3 z - v_1 y^2 - v_2 k^2 - v_3 z^2 \\ &= v_1 y(1-y) + v_2 k(1-k) + v_3 z(1-z) \end{aligned}$$

Also

$$p_{11} - p_{111} = v_1 y^2(1-y) + v_2 k^2(1-k) + v_3 z^2(1-z)$$

and so on. Since the determinant of a product of non-singular matrices equals the product of their determinants, we have the determinant of the matrix in (2.41) equal to the product of the determinants of the two matrices on the right hand side of (2.40). Consider the determinant of the second matrix on the right hand side of (2.40). It may be expanded by elements of the last column, and equals

$$-1/v_2 \begin{vmatrix} 1 & y(1-y) \\ 1 & z(1-z) \end{vmatrix}$$

Since $y = 0.5 + a$ and $z = 0.5 - a$, this is

$$-1/v_2 \begin{vmatrix} 1 & (0.5+a)(0.5-a) \\ 1 & (0.5-a)(0.5+a) \end{vmatrix} = 0$$

Hence

$$|P| = \begin{vmatrix} 1 & p_1 - p_{11} & 1 \\ p_1 & p_{11} - p_{111} & k \\ p_{11} & p_{111} - p_{1111} & k^2 \end{vmatrix} = 0$$

Expanding by the last column, we get

$$\begin{vmatrix} 1 & p_1 - p_{11} \\ p_1 & p_{11} - p_{111} \end{vmatrix} k^2 - \begin{vmatrix} 1 & p_1 - p_{11} \\ p_{11} & p_{111} - p_{111} \end{vmatrix} k$$

$$+ \begin{vmatrix} p_1 & p_{11} - p_{111} \\ p_{11} & p_{111} - p_{1111} \end{vmatrix} = 0 \qquad (2.42)$$

The determinants are easily computed from the data, and k is found by the quadratic formula. It is necessary to turn to the original equations in order to find the other unknowns, which may easily be done now that k is known. We get, for example,

$$a = \sqrt{\frac{p_{111} - p_{11} + (p_1 - p_{11})k + 0.25(p_1 - k)}{p_1 - k}} \qquad (2.43)$$

The second model we want to outline for four interviews has the following structure:

Latent class	*Proportion in latent class*	*Probability of positive response*			
		Time 1	*Time 2*	*Time 3*	*Time 4*
C_1	v_1	1	1	1	1
C_2	v_2	a	a	a	a
C_3	v_3	$b = 1-a$	$b = 1-a$	$b = 1-a$	$b = 1-a$
C_4	v_4	0	0	0	0

What is the logic for proposing this model? It has four latent classes, two of them with perfect reliability, the other two with complementary reliabilities. There are certain cases where mechanical errors and errors made by the interviewer are known to be virtually zero, because of the procedures employed. The magazine data we cited was such a case. In these cases virtually the entire variability is caused by the respondent. Some of these items refer to issues on which it is known that there are extremists and moderates on both sides of the issue. The classes are thought to be ordered, in other words. It is sometimes a reasonable conjecture that the range of error on the part of the extremists is not sufficiently great for their responses to cross the cutting point of the manifest dichotomy, from which it may be inferred plausibly that their responses do not differ from their true position by more than one class. If, for example, attitudes towards war are considered, it may be thought that respondents who are extremely belligerent, while their manifest responses might indicate a greater or lesser degree of pro-war sentiment, would never go so far as to take an anti-war position. Respondents who are only mildly belligerent, on the other hand, might on occasion give anti-war responses. If five interviews are available, it is of course not necessary to assume that the probability of the extreme positions is zero or one, but with only four interviews for analysis, the assumption of perfect reliability for these positions may be considered a sufficiently close approximation to make the use of the present model worth while.

The following equations result from the present assumptions:

$$p_1 = v_1 + av_2 + bv_3$$

$$p_{11} = v_1 + a^2v_2 + b^2v_3$$

$$p_{111} = v_1 + a^3v_2 + b^3v_3$$

$$p_{1111} = v_1 + a^4v_2 + b^4v_3$$

Hence

$$p_1 - p_{11} = a(1-a)v_2 + b(1-b)v_3$$

$$= a(1-a)v_2 + a(1-a)v_3$$

$$a^2(p_1 - p_{11}) = a^3(1-a)v_2 + a^3(1-a)v_3$$

Similarly,

$$a(p_{11} - p_{111}) = a^3(1-a)v_2 + a^2(1-a)^2v_3$$

and

$$p_{111} - p_{1111} = a^3(1-a)v_2 + a(1-a)^3v_3$$

Dividing the difference of the second and third of these equations by the difference of the first and second,

$$\frac{-p_{1111}+p_{111}-ap_{11}+ap_{111}}{ap_{11}-ap_{111}-a^2p_1+a^2p_{11}}=\frac{(1-a-a)a(1-a)^2v_3}{(1-a-a)a^2(1-a)v_3}$$

$$=\frac{1-a}{a}$$

Clearing and collecting terms,

$$a^2(p_1-p_{11})-a(p_1-p_{11})+p_{11}-2p_{111}+p_{1111}=0$$

$$a=\frac{1}{2}+\sqrt{0.25-\frac{p_{11}-2p_{111}+p_{1111}}{p_1-p_{11}}} \tag{2.50}$$

$$b=1-a=\frac{1}{2}-\sqrt{0.25-\frac{p_{11}-2p_{111}+p_{1111}}{p_1-p_{11}}} \tag{2.51}$$

Conditions for Reducibility to the Two-Class Case

If there are four or more interviews available, we may test the assumption that there are only two latent classes with no restrictions placed on the latent probabilities. In addition to the assumed condition that the manifest frequencies of any given order must all be the same, we now have further conditions on the data. Consider the following sequences of proportions:

$$\begin{array}{lllll} 1 & p_1 & p_{11} & p_{111} & \cdots \\ p_1 & p_{11} & p_{111} & p_{1111} & \cdots \end{array}$$

We may form determinants by selecting any square set of four of these proportions which are adjacent in these sequences (see Lazarsfeld and Dudman, 1951). Selecting the first of such determinants, and assuming that there are only two true classes, we have the following relations:

$$\begin{vmatrix} 1 & p_1 \\ p_1 & p_{11} \end{vmatrix} = \begin{vmatrix} \sum_i v_i & \sum_i a_i v_i \\ \sum_i a_i v_i & \sum_i a_i^2 v_i \end{vmatrix} = \begin{vmatrix} v_1 & v_2 \\ a_1 v_1 & a_2 v_2 \end{vmatrix} \begin{vmatrix} 1 & a_1 \\ 1 & a_2 \end{vmatrix}$$

$$= v_1 v_2 \begin{vmatrix} 1 & 1 \\ a_1 & a_2 \end{vmatrix} \begin{vmatrix} 1 & a_1 \\ 1 & a_2 \end{vmatrix}$$

Applying the same analysis to successive determinants of the sequence, we find that

$$\begin{vmatrix} p_1 & p_{11} \\ p_{11} & p_{111} \end{vmatrix} = a_1 a_2 v_1 v_2 \begin{vmatrix} 1 & 1 \\ a_1 & a_2 \end{vmatrix} \begin{vmatrix} 1 & a_1 \\ 1 & a_2 \end{vmatrix}$$

$$\begin{vmatrix} p_{11} & p_{111} \\ p_{111} & p_{1111} \end{vmatrix} = a_1^2 a_2^2 v_1 v_2 \begin{vmatrix} 1 & 1 \\ a_1 & a_2 \end{vmatrix} \begin{vmatrix} 1 & a_1 \\ 1 & a_2 \end{vmatrix}$$

and so on.

From these equations the relations between the determinants may be obtained:

$$\frac{\begin{vmatrix} 1 & p_1 \\ p_1 & p_{11} \end{vmatrix}}{\begin{vmatrix} p_1 & p_{11} \\ p_{11} & p_{111} \end{vmatrix}} = \frac{\begin{vmatrix} p_1 & p_{11} \\ p_{11} & p_{111} \end{vmatrix}}{\begin{vmatrix} p_{11} & p_{111} \\ p_{111} & p_{1111} \end{vmatrix}} = \frac{\begin{vmatrix} p_{11} & p_{111} \\ p_{111} & p_{1111} \end{vmatrix}}{\begin{vmatrix} p_{111} & p_{1111} \\ p_{1111} & p_{11111} \end{vmatrix}} = \ldots \qquad (2.52)$$

These are necessary conditions for the reduction of the data to the two-class case.

The Use of More than Four Interviews

No models will be proposed here for the analysis of data based on more than four interviews. It will be seldom that such data are available for an item such that the assumption of no latent change is reasonable. If such data are available, any model may be used which has fewer unknowns than the number of interviews, or the same number. Since the sum of the proportions in all latent classes is always one, there are $m - 1$ unknown true proportions to be determined if there are m latent classes. In addition, there are m

probabilities. Hence a solution can always be obtained if

$$2m - 1 \leqslant n$$

If no restrictions are put on the latent classes and the latent probabilities, the maximum number of latent classes that can be obtained is consequently

$$m \leqslant \frac{n+1}{2}$$

If there is some reason to assume that the probabilities have relations among themselves which can be specified in advance, then in general one additional class may be found for every two such relations assumed. Of course, the existence of a certain number of latent classes is in itself an assumption. It is usually preferable, if possible, to employ models simpler than the data will permit, in order to have some degrees of freedom left for testing the applicability of the model. In the absence of such testing, considerable caution must be exercised in applying the models which have been proposed in this section. We turn presently to more generally applicable models where latent change is allowed.

Analogue for Continuous Variables

To handle continuous variables, a slightly different set of concepts and notation are required. Since there is an infinite number of latent and manifest classes, we cannot speak of the probability that a person in a latent class gives a certain manifest response. We could, of course, divide the continua into intervals and treat the data as before, but here we wish to deal with the continuous variables as such.

Accordingly, we adopt the standard concepts and notations of test psychology (see Gulliksen, 1950, and Torgerson, 1958). The three basic concepts are "observed score", "true score", and "error". The observed score corresponds to the manifest class for discrete variables, the true score corresponds to the true class, and there is nothing to which the concept of error closely corresponds. Let

X_i = observed score (or value) of ith person on the variable
T_i = true score of ith person on the variable
E_i = error of ith person on the variable.

Then the basic assumption of test theory is that

$$X_i = T_i + E_i \tag{2.53}$$

It is further assumed that the errors are random and uncorrelated with anything else (including the true score) or with themselves at any other time, and that they are normally distributed with mean zero.

In test psychology, as in factor analysis and latent structure analysis, and as in the present paper, it is customary to write equations as if the data were infallible, without worrying about sampling considerations (Gulliksen, 1950). Of course, in practice the data will fit imperfectly, depending on the number of cases in the sample. With infallible data, the assumptions above lead to equality of the means of observed scores and true scores (since the mean of errors is zero), or

$$\bar{X} = \bar{T} \tag{2.54}$$

where the bar over a symbol indicates the mean. Accordingly, we may write

$$x_i = t_i + e_i \tag{2.55}$$

where the lower-case letters refer to deviations from the mean rather than raw scores.

Since the error is uncorrelated with the true score, this gives us

$$s_x^2 = s_t^2 + s_e^2 \tag{2.56}$$

That is, the variance of the observed scores is equal to the sum of the variance of the true scores and the error variance.

The reliability of a test r_r is then defined as the ratio of the true variance to the observed variance, or

$$r_r = s_t^2/s_x^2 \tag{2.57}$$

There is one more concept which will be useful, called the index of reliability. It is the correlation r_{xt} between the true and observed scores, and it can be shown (Gulliksen, 1950) that

$$r_{xt} = s_t/s_x \tag{2.58}$$

The index of reliability, which is the correlation between the true and observed scores, roughly corresponds to the latent probabilities, which link the latent and manifest classes.

We need to add to the notation in order to deal with scores at different times. We shall abandon subscripts referring to persons and use the subscripts to refer to time. Thus

X_i is an observed raw score at time i
x_i is an observed deviation score at time i
T_i is a true raw score at time i
t_i is a true deviation score at time i
E_i is an error raw score at time i
e_i is an error deviation score at time i
$r_{X_iX_j}$ is the correlation between raw observed scores at times i and j
$r_{T_iT_j}$ is the correlation between raw true scores at times i and j
$s^2_{T_i}$ is the variance of raw true scores at time i
$s_{X_iT_j}$ is the covariance between raw observed scores at time i and raw true scores at time j

and so on.

We want to see what the correlation matrix of observed scores will look like under assumptions analogous to those of the latent probability models of this chapter. The assumption that there is no latent change means that the true scores do not change, that is, for any given person,

$$T_1 = T_2 = T_3 = \ldots \qquad (2.59)$$

The assumption that the latent probabilities do not change is equivalent to the assumption that the reliability remains constant, so that

$$s^2_{t_1}/s^2_{x_1} = s^2_{t_2}/s^2_{x_2} = \ldots \qquad (2.60)$$

(2.59) implies equality of the true variances at all times, hence that together with (2.60) implies equality of the observed variances at all times. Now the covariance between observed scores at any pair of times i and j is

$$s_{x_ix_j} = \sum_{i,j} x_ix_j = \sum_{i,j} (t_i + e_i)(t_j + e_j) \qquad (2.61)$$

$$= \sum t_it_j + \sum t_ie_j + \sum t_je_i + \sum e_ie_j$$

$$= \sum t_i^2 = s^2_{t_i}$$

by (2.59) and the assumption that the error term is uncorrelated with anything else or with itself at different times. Hence for any i and j

$$r_{x_i x_j} = s_{t_i}^2 / s_{x_i} s_{x_j} = s_{t_i}^2 / s_{x_i}^2 \tag{2.62}$$

By (2.60) this implies that all $r_{x_i x_j}$ are equal, for any i and j, and by (2.57) they are all equal to r_r, the reliability. Hence the correlation matrix has all elements identical:

	time			
	1	2	3	4
time 1	/	r_r	r_r	r_r
2		/	r_r	r_r
3			/	r_r
4				/

It might be noted in passing that it is not really necessary that the true raw scores be the same at each time, in order to derive this matrix; it would be sufficient if the true deviation scores at each time are the same. In other words, if each respondent maintained his same relative position in the distribution, the matrix above would result (provided all variances remained the same) even though the observed mean might shift or show some kind of a trend. This is different from the discrete case, where the marginals must remain constant in the models of this chapter.

CHAPTER 3

Models Combining Fixed Latent Probabilities with Unsystematic Latent Change

The models of the previous chapter could be called "no change" latent probability models because in them unchanging latent probability alone accounts for the manifest data. However, in addition to the latent probability phenomenon, other things may be going on. Changes from one latent or true class to another may be occurring in an unsystematic or a systematic manner. Such changes will be called here "latent" changes, not to be confused with changes in the latent probabilities.

If the latent changes occur in an unsystematic manner, we do not have a model which accounts for the data completely. Nonetheless, latent classes and their associated latent probabilities may often be determined and may explain part of the total manifest change. Under given assumptions, the turnover may be split into that part which can be accounted for the latent probabilities and that part which cannot, which latter part is then residually defined as the latent change.

It is possible that the latent changes themselves occur in a systematic way; for example, if we consider them as estimates of probabilities they may constitute a stationary Markov process of the sort discussed in Chapter 4.

The present chapter is limited to cases where no assumption is made about the systematicity of the latent changes. Certain other assumptions about these changes must be made in order to obtain solutions, of course, and these assumptions are stated in each case. Moreover, in all the models of this chapter it is assumed that the latent probabilities do not change.

A Case Where Latent Change in Only One Direction is Assumed

If there are only two interviews available, it is impossible to obtain a solution if latent change is assumed to occur in both directions between the two classes that can be obtained on the basis of two interviews. In such cases, it is sometimes plausible to assume that latent change occurs only in one direction. Such cases might arise where an experimental effect has been induced, or where because of some striking event or circumstance there is a marked shift of opinion in one direction. Again, certain factual items such as age and education can shift only in one direction, and if a long time interval elapses between interviews this shift might be significant. For such cases the following model is proposed:

Latent class	*Proportion in latent class*	*Probability of positive response*	
		Time 1	*Time 2*
C_{11}	v_{11}	$0.5 + a$	$0.5 + a$
C_{12}	v_{12}	$0.5 + a$	$0.5 - a$
C_{21}	0	$0.5 - a$	$0.5 + a$
C_{22}	v_{22}	$0.5 - a$	$0.5 - a$

At each given time, there are only two latent classes. But if both times are considered together, there are three classes — those who are in position 1 at both times, those who shift from position 1 to position 2, and those who are in position 2 at both times. The probability of positive response is that of the class to which a respondent belongs at the given time, and the latent probabilities are assumed to be complementary for the two classes.

What equations result from this model? Since we no longer assume that there is no latent change, the marginals and joint frequencies of any given order will no longer be the same. Hence we return to our original notation, in which the position of a given subscript denotes the time to which it applies, while its value denotes the position at that time. The manifest probability of positive response at time 1, $p_{1.}$, will be the sum of the products of the true proportions at time 1 and the respective probabilities of giving the positive response of the classes having these proportions.

$$p_{1.} = (0.5 + a)v_{11} + (0.5 + a)v_{12} + (0.5 - a)v_{22} \qquad (3.1)$$

$$= 0.5 + a(v_{11} + v_{12} - v_{22})$$

By time 2 the respondents in class C_{12} have shifted to position 2, so for $p_{.1}$ we have

$$p_{.1} = (0.5 + a)v_{11} + (0.5 - a)v_{12} + (0.5 - a)v_{22} \qquad (3.2)$$

$$= 0.5 + a(v_{11} - v_{12} - v_{22})$$

For the joint positive frequency, p_{11}, those who gave the positive response at both times, we need the product of the latent probability at time 1 and at time 2 for each class, multiplied by the proportion in that class:

$$p_{11} = (0.5 + a)^2 v_{11} + (0.5 + a)(0.5 - a)v_{12} + (0.5 - a)^2 v_{22} \qquad (3.3)$$

$$= 0.25 + a(v_{11} - v_{22}) + a^2(v_{11} - v_{12} + v_{22})$$

Now

$$0.5(p_{1.} + p_{.1}) = 0.5 + a(v_{11} - v_{22})$$

and

$$p_{1.} - p_{.1} = 2av_{12}$$

Hence

$$p_{11} - 0.5(p_{1.} + p_{.1}) + 0.25 = a^2(v_{11} - v_{12} + v_{22})$$

But

$$v_{11} - v_{12} + v_{22} = v_{11} - v_{12} + 1 - v_{11} - v_{12}$$

$$= 1 - 2v_{12}$$

Therefore

$$p_{11} - 0.5(p_{1.} + p_{.1}) + 0.25 = a^2(1 - 2v_{12})$$

$$= a^2 - a(p_{1.} - p_{.1})$$

Collecting terms, we get

$$a^2 + a(p_{.1} - p_{1.}) - p_{11} + 0.5(p_{1.} + p_{.1}) - 0.25 = 0$$

Let us call the latent probability k, where we have

$$k = 0.5 + a$$

Substituting $k-0.5$ for a, and collecting terms, we get

$$k^2-k(1-p_{.1}+p_{1.})+p_{1.}-p_{11}=0 \tag{3.4}$$

from which

$$k=0.5\left(1-p_{.1}+p_{1.}+\sqrt{(1-p_{1.}-p_{.1})^2+4[12]}\right) \tag{3.5}$$

For the latent proportions we have

$$p_{1.}=kv_{1.}+(1-k)(1-v_{1.})$$

or

$$v_{1.}=\frac{p_{1.}+k-1}{2k-1}$$

Similarly,

$$v_{.1}=\frac{p_{.1}+k-1}{2k-1}$$

Now

$$v_{12}=v_{1.}-v_{.1}=\frac{p_{1.}-p_{.1}}{2k-1} \tag{3.6}$$

by the above, and

$$v_{11}=v_{1.}-v_{12}$$

$$v_{22}=1-v_{11}-v_{12}$$

An Example Comparing this Model to an Earlier One

This example is based on data taken from *Experiments on Mass Communication* by Hovland *et al.* (1949) (this example was given in Wiggins, 1955a; Coleman, 1964 has applied a continuous-time latent process model for the same example with good results). Two groups of soldiers were selected for an experiment on the effects of a radio program on their optimism about the war with Japan. The transcription was designed to reduce what was thought to be over-optimism. One group of soldiers heard the experimental transcription; the other did not. Both groups were asked to estimate the remaining length of the war both before and after the period when the program was given. The time interval between the interviews was about one week.

Since the time interval was short, it is fairly reasonable to assume that the control group experienced no true change during the interval. With the experimental group, it seems reasonable to think that there was true change only in the direction induced by the transcription, i.e., in the direction of longer estimates of the remaining duration of the war. This gives us an opportunity to compare the two-class model assuming no latent change with the present model.

The estimates given by the men were coded in half-year intervals. Their distribution is therefore not a dichotomy, but for illustrative purposes we shall divide the estimates into two groups, those from one through three half-years, and those of four or more half-years.

The manifest data, translated into percentages, are shown in Table 6.

TABLE 6

Estimates of length of war before and after a radio transcription

Experimental group (estimates in half-years)					*Control group (estimates in half-years)*				
		time 2					time 2		
		1–3	4	Total			1–3	4	Total
time	1–3	0.39	0.23	0.62	time	1–3	0.59	0.05	0.64
1	4	0.03	0.35	0.38	1	4	0.07	0.29	0.36
	Total	0.42	0.58	1.00		Total	0.66	0.34	1.00

The data accord prima facie with our expectations. When we apply the model of this section to the experimental group data, and the model of Chapter 2 to the data for the control group, we get the following picture of what would have been obtained had the item been perfectly reliable (see Table 7).

The latent probabilities are both high and strikingly similar. We may take the initial latent distribution of the experimental group and its latent probabilities and generate the data that would be expected if there had been no latent change in that group, getting the following table:

		time 2		
		1–3	4	Total
time 1	1–3	0.59	0.03	0.62
	4	0.03	0.35	0.38
	Total	0.62	0.38	1.00

If this regenerated table is compared with the actual manifest data from the control group, it will be observed that it is fairly similar. This increases our confidence in the use of the present model in such situations. It does not, of course, eliminate the necessity for using a control group if it is desired to measure an experimental effect, since a control group might in some cases also show a trend.

TABLE 7

Latent distribution of estimates of length of war

Experimental group (estimates in half-years)					*Control group (estimates in half-years)*				
		time 2					time 2		
		1–3	4	Total			1–3	4	Total
time 1	1–3	0.41	0.22	0.63	time 1	1–3	0.67	0.00	0.67
	4	0.00	0.37	0.37		4	0.00	0.33	0.33
	Total	0.40	0.59	1.00		Total	0.67	0.33	1.00
Latent probability = 0.97					Latent probability = 0.94				

Models Allowing Latent Change among all Classes

Most of the items that will be usefully subjected to the kind of analysis presented in this paper are such that respondents can and do change to a certain extent from any latent class to any other class, in the course of time. The present section treats some of the models which are based on the assumption that respondents do change their latent positions. This will introduce a larger number of unknowns into the models, but it also adds new items of information, for we no longer assume that the manifest frequencies of any given order are all the same. No solutions can be obtained with fewer than three interviews, but on the other hand items allowing free change are more likely to be repeated a number of times than items where the change is restricted, for in panel studies it is often the change which is the object of study.

Three Interviews, Two Latent Classes, Terminal Latent Position Independent of Initial Latent Position, Controlling Middle Latent Position

If we have the responses on a dichotomous item for three interviews, we have eight pieces of information — p_{111}, $p_{112}, \ldots$, p_{222}. If there are two latent classes at any one time and we allow free movement between these classes, we have over the entire time eight latent classes — v_{111}, $v_{112}, \ldots$, v_{222}. In addition, we must find the two probabilities of positive response, which we shall now call a_1 and a_2, referring to latent classes C_1 and C_2 respectively. Evidently we must make further assumptions which will reduce the number of unknowns by two. We might think of placing restrictions on the latent probabilities, but this would reduce the unknowns by only one.

It seems profitable to think in somewhat different terms. We are dealing with a time sequence, and we may consider assumptions which apply to the process through time. It is a reasonable first approximation to assume that the latent changes from time 2 to time 3 are independent of the latent changes from time 1 to time 2. Another way of stating this is that if we want to predict the positions of respondents at time 3, and we know the positions at time 2, our prediction is not improved by knowing also the positions at time 1. In other words, the conditional distribution at time 3 given time 2 is the same as the conditional distribution at time 3 given time 2 and time 1. This is similar to one assumption made in a first-order Markov chain, except that it is on the latent level.

Let us see what restrictions this assumption places upon our latent classes v_{ijk}. It will be recalled that the position of a subscript indicates the time to which it refers, while the subscript "1" refers to class C_1, the subscript "2" refers to class C_2, and the subscript "." refers to the sum of both classes. Thus $v_{.1.}$ means the porportion in latent class C_1 at time 2, summed over all latent classes at time 1 and 3; in other words, the marginal proportion in latent class C_1 at time 2. Now of those in class C_1 at time 2, a certain number will remain in C_1 at time 3. Expressed as a proportion of the total sample, this number will be $v_{.11}$. Expressed as a proportion of those who were in class C_1 at time 2, this number will be

$$\frac{v_{.11}}{v_{.1.}}$$

This ratio shows the probability of going from C_1 at time 2 to C_1 at time 3. Now consider those who are in C_1 at both times 1 and 2. The probability that they will remain in C_1 at time 3 is

$$\frac{v_{111}}{v_{11.}}$$

Similarly, the probability that those who are in class C_2 at time 1 and class C_1 at time 2 will remain in C_1 at time 3 is

$$\frac{v_{211}}{v_{21.}}$$

Now what does our assumption mean? It says that the probability of going from any given latent class at time 2 to any given class at time 3 is independent of the position at time 1. This means, e.g., that

$$\frac{v_{111}}{v_{11.}} = \frac{v_{211}}{v_{21.}}$$

That is, the probability of going from class C_1 at time 2 to class C_1 at time 3 is the same, regardless what the position at time 1 is. The left hand side of the equation above shows the probability of going from C_1 at time 2 to C_1 at time 3 when the respondents were in C_1 at time 1. The right hand side of the equation shows the probability of going from C_1 at time 2 to C_1 at time 3 when the respondents were in C_2 at time 1. By our assumption, these probabilities are the same. Similarly,

$$\frac{v_{112}}{v_{11.}} = \frac{v_{212}}{v_{21.}}$$

That is, the probability of going from C_1 at time 2 to C_2 at time 3 is the same, regardless of the position at time 1. Now if we look at these two equations, we see that they have the same denominator on each side. We can therefore derive from them the equation

$$\frac{v_{111}}{v_{112}} = \frac{v_{211}}{v_{212}}$$

or

$$v_{111}v_{212} = v_{211}v_{112}$$

In determinantal form, this is

$$\begin{vmatrix} v_{111} & v_{112} \\ v_{211} & v_{212} \end{vmatrix} = 0 \tag{3.7}$$

Notice the form of this determinant. The second subscript is the same for all four elements. The first and third subscripts are arranged in the customary manner for determinants. The form is important, for by our assumption, all determinants of this form must vanish.

We have considered only persons in class C_1 at time 2. If we consider those in class C_2 at time 2, then by the same reasoning used before, our assumption implies that

$$\frac{v_{121}}{v_{122}} = \frac{v_{221}}{v_{222}}$$

This says that of those in class C_2 at time 2, the ratio of those going to C_1 at time 3 to those remaining in C_2 at time 3 is the same for both classes to which they might have belonged at time 1. In determinantal form, it is

$$\begin{vmatrix} v_{121} & v_{122} \\ v_{221} & v_{222} \end{vmatrix} = 0$$

This determinant is exactly like the previous one, except that the middle subscript of all four elements is 2 instead of 1.

We thus have two relations, each involving four of the eight ultimate latent class frequencies. There are other relations also implied by our assumption, but they are not independent of those we have already established. Thus our assumption implies two independent restrictions on the unknowns, which enables us to solve the equations from the manifest data.

Now let us consider the following array of manifest data:

$$\begin{matrix} 1 & p_{..1} \\ p_{1..} & p_{1.1} \\ p_{.1.} & p_{.11} \\ p_{11.} & p_{111} \end{matrix}$$

This array contains all the information in the data. Certain symmetries in the array may be noted. The square at the top does not contain any reference to time 2, while every element of the bottom square does refer to time 2. With that one difference, the two squares are alike. Let us call the first interview the "initial" interview and the third interview the "terminal" interview. The top square contains all the information about this pair of interviews alone, including the marginal frequency at each time and the joint frequency at both times. The bottom square is the same thing but stratified, so to speak, by the positive response at time 2.

Now let us express the matrix of the top square in terms of the latent structure. The first element, 1, will equal the sum of all eight latent frequencies. We also have

$$p_{..1} = (v_{111} + v_{211})a_1 + (v_{112} + v_{212})a_2 + (v_{121} + v_{221})a_1 + (v_{122} + v_{222})a_2$$

In other words, only the position at time 3 matters, so every ultimate latent frequency with the subscript 1 in the third position is multiplied by a_1 and every ultimate true frequency with the subscript 2 in the third position is multiplied by a_2. We have set the equation up in its present form for a reason which will become apparent. In terms of vectors and matrices, the equation would be

$$p_{..1} = (1 \quad 1) \begin{pmatrix} v_{111} & v_{112} & v_{121} & v_{122} \\ v_{211} & v_{212} & v_{221} & v_{222} \end{pmatrix} \begin{pmatrix} a_1 \\ a_2 \\ a_1 \\ a_2 \end{pmatrix}$$

Call the matrix in $v_{ijk} V$. Then V premultiplied by the vector $(1 \quad 1)$ becomes a row vector with each element the sum of the elements of the corresponding column of V. The final vector then takes care of the probabilities for time 3, as seen in the previous equation.

Setting up $p_{1..}$ in matrix notation,

$$p_{1..} = (a_1 \quad a_2) \begin{pmatrix} v_{111} & v_{112} & v_{121} & v_{122} \\ v_{211} & v_{212} & v_{221} & v_{222} \end{pmatrix} \begin{pmatrix} 1 \\ 1 \\ 1 \\ 1 \end{pmatrix}$$

For $p_{1..}$, the only subscript of v_{ijk} we have to worry about is the first. The vector $(a_1 \; a_2)$ takes care of this subscript for each column of V. The final vector merely sums the result.

Finally,

$$p_{1.1} = (a_1 \quad a_2) \begin{pmatrix} v_{111} & v_{112} & v_{121} & v_{122} \\ v_{211} & v_{212} & v_{221} & v_{222} \end{pmatrix} \begin{pmatrix} a_1 \\ a_2 \\ a_1 \\ a_2 \end{pmatrix}$$

That is, in order to get $p_{1.1}$, we have to multiply any given true ultimate frequency v_{ijk} by a_i, the probability referring to time 1, and by a_k, the probability referring to time 3.

Combining the equations, we get

$$\begin{pmatrix} 1 & p_{..1} \\ p_{1..} & p_{1.1} \end{pmatrix} = \begin{pmatrix} 1 & 1 \\ a_1 & a_2 \end{pmatrix} \begin{pmatrix} v_{111} & v_{112} & v_{121} & v_{122} \\ v_{211} & v_{212} & v_{221} & v_{222} \end{pmatrix} \begin{pmatrix} 1 & a_1 \\ 1 & a_2 \\ 1 & a_1 \\ 1 & a_2 \end{pmatrix}$$

We thus have the matrix of the top square of manifest data expressed as the product of a square matrix and two rectangular matrices. The product of the two rectangular matrices alone will be a square matrix of order 2. By a well known theorem of determinants we have

$$\begin{vmatrix} 1 & p_{..1} \\ p_{1..} & p_{1.1} \end{vmatrix} = \begin{vmatrix} 1 & 1 \\ a_1 & a_2 \end{vmatrix} \left| \begin{pmatrix} v_{111} & v_{112} & v_{121} & v_{122} \\ v_{211} & v_{212} & v_{221} & v_{222} \end{pmatrix} \begin{pmatrix} 1 & a_1 \\ 1 & a_2 \\ 1 & a_1 \\ 1 & a_2 \end{pmatrix} \right| \tag{3.9}$$

The determinant on the right hand side of this equation is the determinant of a product of rectangular matrices. By the theorem of multiplication of determinantal arrays (see Aitken, 1948), this determinant is equal to the sum of six products made by pairing each minor of order 2 from 2 columns of the first rectangular matrix, V, with the minor of order 2 from the corresponding rows of the second rectangular matrix (in a_i), which we shall call A. This theorem is essential to the solution which follows.

We next consider the determinant of the matrix which is the bottom square of the array with which we began. It is

$$\begin{vmatrix} p_{.1.} & p_{.11} \\ p_{11.} & p_{111} \end{vmatrix} = \begin{vmatrix} 1 & 1 \\ a_1 & a_2 \end{vmatrix} \left| \begin{pmatrix} v_{111} & v_{112} & v_{121} & v_{122} \\ v_{211} & v_{212} & v_{221} & v_{222} \end{pmatrix} \begin{pmatrix} a_1 & a_1 \cdot a_1 \\ a_1 & a_1 \cdot a_2 \\ a_2 & a_2 \cdot a_1 \\ a_2 & a_2 \cdot a_2 \end{pmatrix} \right| \tag{3.10}$$

Now compare (3.10) with (3.9). The introduction of time 2 as a stratifier has had the effect of multiplying the first row of the A matrix by a_1, the second row by a_1, the third row by a_2, and the fourth row by a_2. Consider why this is so. The composition of the V matrix is such that all the elements of any given column of V refer to the same latent classes at times 2 and 3, while all the elements of the first row refer to latent class C_1 at time 1, and all the elements of the second row refer to latent class C_2 at time 1. The columns are arranged in lexical order with respect to times 2 and 3. In the matrix equation which corresponds to the determinantal equation of (3.9), the first matrix on the right hand side takes care of time 1 and the last matrix on the right hand side takes care of time 3. The equation (3.10) merely adds the stratifier, time 2. Now whenever a column of V refers to the latent class C_1 at time 2, the corresponding row of A is multiplied by a_1, and wherever a column of V refers to the latent class C_2 at time 2, the corresponding row of A is multiplied by a_2. Let us call the resulting matrix A_1.

To evaluate the determinant on the right hand side of (3.10) which is the product of V and A_1, we must take the sum of the products of all the minors of order 2 from columns of the V matrix with the minors of order 2 from the corresponding row of A_1. But our hypothesis, expressed in (3.7) and (3.8), is that all the minors of the latent proportions v_{ijk} of order 2 which refer to the same latent class at time 2 and which for times 1 and 3 are arranged in the customary manner for matrices vanish. This means that in evaluating the determinant of VA_1, all the terms of the sum which contain determinants chosen from two rows of A_1 which are actually two rows of A both multiplied by the same unknown, either a_1 or a_2, will vanish. Furthermore, since V is the same in both (3.10) and (3.9), the corresponding terms in the sum which is the evaluated determinant of VA will also vanish. The only terms of A_1 which will not vanish for this reason are those which contain determinants chosen from two rows which are rows of A one of which is multiplied by a_1 and one of which is multiplied by a_2.

Hence every term in the evaluated determinant of VA_1 will be identical with the corresponding term in the evaluated determinant of VA, except that every determinant from A_1 will be a determinant of A with one row multiplied by a_1 and the other row multiplied by a_2. But this is the same as multiplying every determinant by $a_1 a_2$. Hence, finally, the evaluated product on the right hand side of (3.10) will be identical with the evaluated product on the right hand side of (3.9) multiplied by $a_1 a_2$, or

$$\begin{vmatrix} p_{.1.} & p_{.11} \\ p_{11.} & p_{111} \end{vmatrix} = a_1 a_2 \begin{vmatrix} 1 & p_{..1} \\ p_{1..} & p_{1.1} \end{vmatrix} \tag{3.11}$$

We next consider the determinants of the two matrices which can be formed by substituting one column of the array in the bottom square of the initial array for the corresponding column of the array in the top square, namely,

$$\begin{vmatrix} 1 & p_{.11} \\ p_{1..} & p_{111} \end{vmatrix} \quad \text{and} \quad \begin{vmatrix} p_{.1.} & p_{..1} \\ p_{11.} & p_{1.1} \end{vmatrix}$$

For these determinants we get the following equations:

$$\begin{vmatrix} 1 & p_{.11} \\ p_{1..} & p_{111} \end{vmatrix} = \begin{vmatrix} 1 & 1 \\ a_1 & a_2 \end{vmatrix} \begin{vmatrix} \begin{pmatrix} v_{111} & v_{112} & v_{121} & v_{122} \\ v_{211} & v_{212} & v_{221} & v_{222} \end{pmatrix} \begin{pmatrix} 1 & a_1 \cdot a_1 \\ 1 & a_1 \cdot a_2 \\ 1 & a_2 \cdot a_1 \\ 1 & a_2 \cdot a_2 \end{pmatrix} \end{vmatrix} \tag{3.12}$$

$$\begin{vmatrix} p_{.1.} & p_{..1} \\ p_{11.} & p_{1.1} \end{vmatrix} = \begin{vmatrix} 1 & 1 \\ a_1 & a_2 \end{vmatrix} \left| \begin{pmatrix} v_{111} & v_{112} & v_{121} & v_{122} \\ v_{211} & v_{212} & v_{221} & v_{222} \end{pmatrix} \begin{pmatrix} a_1 & a_1 \\ a_1 & a_2 \\ a_2 & a_1 \\ a_2 & a_2 \end{pmatrix} \right| \tag{3.13}$$

On the right hand side of these equations, the first determinant and the first rectangular matrix V are the same as before. In (3.12), the second rectangular matrix is the matrix A with the rows of the second column multiplied by a_1 or a_2 accordingly as the corresponding column of V refers to class C_1 or class C_2 at time 2. The second rectangular matrix of (3.13) is the matrix A with the rows of the first column multiplied by a_1 or a_2 in the same manner. In both cases, when evaluating the determinant of the product matrix, the terms of the respective summations which have determinants which involve two rows from A with either the first (in the case of (3.13)) or last (in the case of (3.12)) element of both rows multiplied by the same unknown, a_1 or a_2, will vanish, because these determinants from A will be multiplied by determinants from V which vanish by hypothesis. The sum of the determinants on the left hand side of (3.12) and (3.13) will therefore be equal to the product of $\begin{vmatrix} 1 & 1 \\ a_1 & a_2 \end{vmatrix}$ and a sum of products of the following type:

$$\begin{vmatrix} v_{1_{im}} & v_{1_{jn}} \\ v_{2_{im}} & v_{2_{jn}} \end{vmatrix} \left\{ \begin{vmatrix} 1 & a_i \cdot a_m \\ 1 & a_j \cdot a_n \end{vmatrix} + \begin{vmatrix} a_i & a_m \\ a_j & a_n \end{vmatrix} \right\} \quad i \neq j$$

Evaluating the sum of two determinants in a,

$$\begin{aligned} a_j a_n - a_i a_m + a_i a_n - a_j a_m &= (a_i + a_j)(a_n - a_m) \\ &= (a_i + a_j) \begin{vmatrix} 1 & a_m \\ 1 & a_n \end{vmatrix} \end{aligned}$$

Since $i \neq j$,

$$a_i + a_j = a_1 + a_2$$

for any i and j. Thus this term can be factored out of each term in the expansion, and what remains is simply the expansion of VA. Therefore

$$\begin{vmatrix} 1 & p_{.11} \\ p_{1..} & p_{111} \end{vmatrix} + \begin{vmatrix} p_{.1.} & p_{..1} \\ p_{11.} & p_{1.1} \end{vmatrix} = (a_1 + a_2) \begin{vmatrix} 1 & p_{..1} \\ p_{1..} & p_{1.1} \end{vmatrix} \tag{3.14}$$

In order to find a_1 and a_2, we merely have to solve the quadratic equation

$$x^2 - \frac{\begin{vmatrix} 1 & p_{.11} \\ p_{1..} & p_{111} \end{vmatrix} + \begin{vmatrix} p_{.1.} & p_{..1} \\ p_{11.} & p_{1.1} \end{vmatrix}}{\begin{vmatrix} 1 & p_{..1} \\ p_{1..} & p_{1.1} \end{vmatrix}} x + \frac{\begin{vmatrix} p_{.1.} & p_{.11} \\ p_{11.} & p_{111} \end{vmatrix}}{\begin{vmatrix} 1 & p_{..1} \\ p_{1..} & p_{1.1} \end{vmatrix}} = 0 \tag{3.15}$$

When the two roots are found, the larger root is designated a_1 and the smaller root a_2. Now we wish to find the relative frequencies in the latent classes. We have equations of the following types:

$$p_{1..} = a_1 v_{1..} + a_2 v_{2..}$$

$$p_{11.} = a_1^2 v_{11.} + a_1 a_2 (v_{12.} + v_{21.}) + a_2^2 v_{22.}$$

From the first type we can derive $v_{1..}$ etc. as follows:

$$\begin{aligned} p_{1..} &= a_1 v_{1..} + a_2 v_{2..} \\ &= a_1 v_{1..} + a_2 (1 - v_{1..}) \\ &= a_2 + (a_1 - a_2) v_{1..} \end{aligned}$$

$$v_{1..} = \frac{p_{1..} - a_2}{(a_1 - a_2)} \tag{3.16a}$$

Similarly,

$$v_{.1.} = \frac{p_{.1.} - a_2}{(a_1 - a_2)} \tag{3.16b}$$

and

$$v_{..1} = \frac{p_{..1} - a_2}{a_1 - a_2} \tag{3.16c}$$

From equations of the second type we can derive $v_{11.}$ etc. as follows:

$$p_{11.} = a_1^2 v_{11.} + a_1 a_2 (v_{12.} + v_{21.}) + a_2^2 v_{22.}$$

$$= a_1^2 v_{11.} + a_1 a_2 (v_{1..} - v_{11.} + v_{.1.} - v_{11.})$$

$$+ a_2^2 (1 - v_{1..} - v_{.1.} + v_{11.})$$

$$= (a_1^2 - 2a_1 a_2 + a_2^2) v_{11.} + a_2(a_1 - a_2)(v_{1..} + v_{.1.}) + a_2^2 (a_1 - a_2)^2 v$$

$$(a_1 - a_2)^2 v_{11.} = p_{11.} - a_2(a_1 - a_2)(v_{1..} + v_{.1.}) - a_2^2$$

Substituting for $v_{1..}$ and $v_{.1.}$ from (3.16a) and (3.16b),

$$(a_1 - a_2)^2 v_{11.} = p_{11.} - a_2(a_1 - a_2) \frac{(p_{1..} - a_2 + p_{.1.} - a_2)}{(a_1 - a_2)} - a_2^2$$

$$= p_{11.} - a_2(p_{1..} + p_{.1.}) + a_2^2$$

or

$$v_{11.} = \frac{p_{11.} - a_2(p_{1..} + p_{.1.}) + a_2^2}{(a_1 - a_2)^2} \tag{3.16d}$$

Similarly

$$v_{1.1} = \frac{p_{1.1} - a_2(p_{1..} + p_{..1}) + a_2^2}{(a_1 - a_2)^2} \tag{3.16e}$$

and

$$v_{.11} = \frac{p_{.11} - a_2(p_{.1.} + p_{..1}) + a_2^2}{(a_1 - a_2)^2} \tag{3.16f}$$

In order to find v_{111}, we go back to the restriction imposed on the latent frequencies by the assumption that the latent position at time 3 is independent of the latent position at time 1 (controlling latent position at time 2). This gave us equation (3.7), which was

$$\begin{vmatrix} v_{111} & v_{112} \\ v_{211} & v_{212} \end{vmatrix} = 0$$

Adding the first row of this determinant to the second, and then adding the new first column to the second,

$$\begin{vmatrix} v_{111} & v_{112} \\ v_{211} & v_{212} \end{vmatrix} = \begin{vmatrix} v_{111} & v_{112} \\ v_{.11} & v_{.12} \end{vmatrix} = \begin{vmatrix} v_{111} & v_{11.} \\ v_{.11} & v_{.1.} \end{vmatrix}$$

Expanding the last of these determinants,

$$v_{111}v_{.1.} - v_{11.}v_{.11} = 0$$

or

$$v_{111} = \frac{v_{11.}v_{.11}}{v_{.1.}} \tag{3.16g}$$

The term on the right hand side can be computed from (3.16d), (3.16f), and (3.16b). From the set (3.16a)—(3.16g) all eight ultimate latent relative frequencies v_{ijk} can be computed, if desired, as well as all the probabilities of latent change.

Examples of the Two Class, Three Interview Model

We shall refer once more to the magazine data we have used before. Even though there is reason to think that a two class model does not fit the data well, we shall use the present model for comparison with previous cases where the same data were used. We shall then introduce a new example, on the basis of which we shall later compare another model to the present one.

The magazine data have been previously reported, and will not be repeated here. When the present model is applied to the data, the following values are obtained for the latent class proportions and latent probabilities.

Magazine	*Latent class*	*Proportion in latent class*			*Latent probability*
		Time 1	*Time 2*	*Time 3*	
"A"	reader	0.075	0.068	0.065	0.829
	non-reader	0.925	0.932	0.935	0.933
"B"	reader	0.129	0.120	0.106	0.669
	non-reader	0.872	0.880	0.894	0.960

The results, previously quoted, of applying the two-class model assuming no latent change to the same data, were as follows:

Magazine	*Latent class*	*Proportion in latent class*	*Latent probability*
"A"	reader	0.057	0.745
	non-reader	0.943	0.918
"B"	reader	0.119	0.605
	non-reader	0.881	0.952

Comparing the reliabilities, we see that the assumption that there was no latent change, which was probably false, had the effect of reducing the latent probabilities obtained, especially for latent readers. Both magazines, but especially magazine "B", suffered a steady decline in the number of latent readers during the period covered by the interviews. In the earlier model the latent changes were included in the random turnover, hence the latent probability seemed lower than it actually is.

The second example we shall give is taken from the voting study conducted in Erie County, Ohio, in 1940. See Lazarsfeld *et al.* (1948) for a fuller description of this study. The data used for present purposes were not presented in full in that volume. A random sample of some six hundred adults was selected, and this sample was interviewed seven times at intervals of about a month, beginning in May. The following question was repeated on all six interviews taken before the election: "Regardless of which man or party you would like to see elected, which party do you think actually will be elected?" Our data refer to four hundred and thirty respondents who answered this question on all six interviews. Most of the responses fell into the categories "Democratic," "Republican," and "Don't Know." For present purposes respondents have been grouped into "Democratic" and "Not Democratic."

The advantage of having six interviews, during which considerable net changes of opinion occurred, is that some inferences can be made about the applicability of the model. If the assumption that the latent item structure remains the same over every successive period of three interviews is valid, then the latent probabilities should remain constant over the entire period of six interviews. In addition, from each successive set of three interviews the latent marginals for each time involved can be computed, and since the sets of three overlap, some of the latent marginals should be the same.

TABLE 8

Latent probabilities for "winner expectation," Erie County, 1940

Interviews used	*May—June —July*	*June—July —Aug.*	*July—Aug. —Sept.*	*Aug.—Sept. —Oct.*
Democrats	0.97	0.85	0.96	0.95
Not Democrats	0.97	0.99	0.91	0.95

When the computations for the present model are made for each successive set of three interviews, the following latent probabilities are obtained (see Table 8).

With the exception of the second set of interviews, the latent probabilities show considerable stability. They are, in fact, very high — suspiciously high — ranging from 0.95 to 0.99 for both classes and for all successive sets, with two exceptions. The latent probability for latent Democrats computed from June—July—August interviews fell to 0.85, and the latent probability for latent non-Democrats computed from July—August—September fell to 0.91.

Analysis of the reason for these drops helps to clarify the question of the applicability of the model, and also gives some insight into the substantive situation. The Republican convention occurred between the June and July interviews. A strong candidate, Willkie, was nominated, and the manifest data show at this time a strong trend in expectations away from the Democrats. It must be kept in mind that for the present analysis the "Don't Know" or doubtful respondents are classified with the Republicans. It is reasonable to think that many respondents who previously had truly expected the Democrats to win now shifted their latent opinions to the "Don't Know" or "Republican" categories, on the basis of the Republican nomination of a strong candidate. The Democratic convention then took place between the July and August interviews, and Roosevelt, a very strong candidate, was nominated. With this event, the balance returned in a sense to the status quo ante, and many respondents changed to Democratic expectations. It is reasonable to suppose that under the impact of this event those respondents who had previously shifted from a democratic expectation to a Republican or doubtful expectation were more likely to shift back to Democratic than were those who had entertained a Republican or doubtful expectation from the first. Let us call this process "reversion," by which we mean that respondents who have held a position and shifted away from it are more likely to change back to that position than respondents who have never held the position before are to change to it. (Vote intention also displays this process, as shown in an unpublished memorandum on the Erie County study by Rowena Wyant.)

This process of reversion has a definite relation to the kind of mathematical model we have employed. Suppose we have three interviews of a panel on an item such that no latent change has occurred, with m latent classes, $v_1, v_2, \ldots, v_m$, and their associated probabilities $a_1,a_2,\ldots,a_m$. For the first two interviews, the manifest data can show change in two ways, as revealed in $p_{12.}$ and $p_{21.}$. Taking the third interview into account also, we have

$$p_{12.} = p_{121} + p_{122}$$

and

$$p_{21.} = p_{211} + p_{212}$$

Now our fundamental assumptions give us the following equations:

$$\begin{aligned} p_{121} &= a_1^2(1-a_1)v_1 + a_2^2(1-a_2)v_2 + \ldots + a_m^2(1-a_m)v_m \\ &= p_{211} = p_{112} \\ p_{212} &= a_1(1-a_1)^2v_1 + a_2(1-a_2)^2v_2 + \ldots + a_m(1-a_m)^2v_m \\ &= p_{122} = p_{221} \end{aligned}$$

Hence

$$\begin{aligned} p_{12.} + p_{21.} &= p_{121} + p_{122} + p_{211} + p_{212} \\ &= 2p_{121} + 2p_{212} \end{aligned}$$

or

$$p_{121} = p_{212} = 0.5(p_{12.} + p_{21.})$$

This means that on the basis of latent probabilities alone *half* of the respondents who according to manifest data change their position from the first interview to the second will on the third interview revert to their original positions. If we consider those who do not show manifest change between the first two interviews, we have

$$p_{11.} = p_{111} + p_{112}$$

and

$$p_{22.} = p_{221} + p_{222}$$

But since, in general (even with fallible data), $p_{111} > p_{112}$ and $p_{222} > p_{221}$, less than half of those who do not show manifest change between the first two interviews will show change from the second to the third interview. In other words, in the manifest data, the changers have a greater chance of reverting than the nonchangers have of changing on the third interview, on the basis of latent probability alone.

This fact has a double implication. The implication for manifest data is that with classes with latent probabilities less than one (which includes virtually all items used) the manifest data from a panel study will generally show that changers are more likely to revert than nonchangers are to change to a given position at a given time. This is a statistical artifact, and does not necessarily mean that there is a process of real reversion. Whether there is such a process can only be determined by computing the latent probability of the item and making the kind of analysis we have made here.

The other implication, which concerns us at present, is that a process of real reversion can affect our estimates of latent probability. Since on the basis of latent probability alone half the manifest changers will revert, our models use the amount of manifest reversion to estimate the amount of random change. Now if there is a true process of reversion, these models will necessarily overestimate the amount of random change, which means that they will underestimate the latent probability of the item.

This is precisely what has happened to the latent probability for the true Democrats in our computations for the June—July—August data. The reason we know this is that we are able to make a substantive analysis which supports the hypothesis of real (or latent) reversion, and also because the latent probability for the same class is stable and consistently higher for all the other three sets of interviews. But the occurrence of real reversion implies that our model does not apply in this situation, for true reversion means that changes from time 2 to time 3 are not independent of changes from time 1 to time 2, and our model assumes that there is such independence. If, however, the assumption fails for the June—July—August set of interviews, there is no reason to regard it as valid for any of the other sets, since the same item is involved. We need to weaken our assumptions.

There are two lines we might follow in doing this. We might continue to assume that there are only two latent classes at any given time, and that the latent changes over a given interval are independent, not of the latent changes over the previous interval, but of the latent changes over some earlier interval. Thus if we have four interviews we could assume that to predict the latent positions at time 4 only time 3 and time 2 are required, a knowledge of time 1 latent position adding nothing to the prediction. This is the case we shall take up in the next section. On the other hand, we could assume that there are more than two classes in the structure. This might take care of the different probabilities of latent change for the previous changers and non-changers, or it might not, depending on the situation.

Speaking practically, the second line of development mentioned has little utility in many fields of application when dichotomous items are being used, because too many interviews would be required to find more than two latent classes when latent change is allowed. If items with multiple response categories are used, however, it is feasible to compute a larger number of latent classes with a given number of interviews.

Four Interviews, Two Latent Classes, Latent Terminal Position Independent of Latent Initial Position, Controlling Latent Positions at Second and Third Times

The notation is the same as before, except that we now have four subscripts. We have $2^4 = 16$ pieces of information and $2^4 = 16$ ultimate latent relative frequencies v_{hijk}. We are assuming that if the latent class at time 2 and time 3 is given, the latent position at time 4 is independent of the latent position at time 1. Another way of expressing this is to say that the probability of going to a given class at time 4 depends only on times 2 and 3, and not on time 1. This means that if we let the first subscripts f and g denote the two latent positions at time 1,

$$\frac{v_{fij1}}{v_{fij.}} = \frac{v_{gij1}}{v_{gij.}}$$

and

$$\frac{v_{fij2}}{v_{fij.}} = \frac{v_{gij2}}{v_{gij.}}$$

Combining these equations, and letting $f = 1$ and $g = 2$, we get

$$\frac{v_{1ij1}}{v_{1ij2}} = \frac{v_{2ij1}}{v_{2ij2}}$$

for all i and j. In determinantal form, this yields the following four independent restrictions on the ultimate latent frequencies:

$$\begin{vmatrix} v_{1111} & v_{1112} \\ v_{2111} & v_{2112} \end{vmatrix} = 0 \qquad (3.17)$$

$$\begin{vmatrix} v_{1121} & v_{1122} \\ v_{2121} & v_{2122} \end{vmatrix} = 0 \tag{3.18}$$

$$\begin{vmatrix} v_{1211} & v_{1212} \\ v_{2211} & v_{2212} \end{vmatrix} = 0 \tag{3.19}$$

$$\begin{vmatrix} v_{1221} & v_{1222} \\ v_{2221} & v_{2222} \end{vmatrix} = 0 \tag{3.20}$$

Now consider the following array of manifest data:

(1)		(2)		(3)		(4)	
1	$p_{...1}$	$p_{.1..}$	$p_{.1.1}$	$p_{..1.}$	$p_{..11}$	$p_{.11.}$	$p_{.111}$
$p_{1...}$	$p_{1..1}$	$p_{11..}$	$p_{11.1}$	$p_{1.1.}$	$p_{1.11}$	$p_{111.}$	p_{1111}

The numbers in parentheses are merely for identification of the square arrays. The first square contains all the information about the pair of initial and terminal interviews, with no information about the intervening interviews. Let us call the determinant of this square the "2-positive" determinant. This name expresses the fact that each term in the expansion of this determinant will be a product having in its subscripts a reference to the positive response, or response 1, at two interviews, namely, the first and last. The second square is the first stratified by time 2; the third square is the first stratified by time 3; and the fourth square is the first stratified by times 2 and 3. The whole array contains all the information available.

Now let us set up an equation for the matrix of (1) in terms of the latent structure, in exactly the same manner as we did for the 2-positive matrix in the previous model. We get

$$\begin{pmatrix} 1 & p_{...1} \\ p_{1...} & p_{1..1} \end{pmatrix} = \begin{pmatrix} 1 & 1 \\ a_1 & a_2 \end{pmatrix}$$

$$\cdot \begin{pmatrix} v_{1111} & v_{1112} & v_{1121} & v_{1122} & v_{1211} & v_{1212} & v_{1221} & v_{1222} \\ v_{2111} & v_{2112} & v_{2121} & v_{2122} & v_{2211} & v_{2212} & v_{2221} & v_{2222} \end{pmatrix}$$

$$\cdot \begin{pmatrix} 1 & a_1 \\ 1 & a_2 \\ 1 & a_1 \\ 1 & a_2 \\ 1 & a_1 \\ 1 & a_2 \\ 1 & a_1 \\ 1 & a_2 \end{pmatrix} \tag{3.21}$$

Let us call the first matrix on the right hand side of (3.21) $\bar{A}'$, the second matrix V, and the third matrix A. Notice how the elements of V are arranged. Every element of the first row of V refers to position 1 at time 1; every element of the second row refers to position 2 at time 1. The columns are arranged in lexical order with respect to the second, third, and fourth subscripts. This has the effect of alternating positions 1 and 2 at time 4 in the successive columns. The matrix $\bar{A}'$ takes care of time 1, while A takes care of time 4, as regards the probabilities of positive response for the latent classes. The positions at times 2 and 3 do not matter in the 2-positive matrix. Equation (3.21) can be put in determinantal form, as follows:

$$\begin{vmatrix} 1 & p_{...1} \\ p_{1...} & p_{1..1} \end{vmatrix} = |\bar{A}'| \quad |VA| \tag{3.22}$$

V and A are rectangular matrices, and the determinant of their product can be evaluated by the theorem of multiplication of determinantal arrays as the sum of $(8.7)/2 = 28$ products of minors of order 2 chosen from 2 columns of V and minors of order 2 chosen from the corresponding rows of A. By (3.17)—(3.20), each term of this expansion which contains a minor of V with the same second and third subscripts in both columns will vanish. In particular, the terms containing

$$\begin{vmatrix} v_{1111} & v_{1112} \\ v_{2111} & v_{2112} \end{vmatrix}$$

and

$$\begin{vmatrix} v_{1221} & v_{1222} \\ v_{2221} & v_{2222} \end{vmatrix}$$

will vanish. The remaining terms will have the following form:

$$\begin{vmatrix} v_{1hir} & v_{1jks} \\ v_{2hir} & v_{2jks} \end{vmatrix} \begin{vmatrix} 1 & a_r \\ 1 & a_s \end{vmatrix}$$

where h, i, j, and k cannot all be either 1 or 2. (There are other limitations on the subscripts, but they are omitted as irrelevant for present purposes.) Considering the second and third subscripts only, these remaining terms will be of the three following types:

(1) h, i, j, k is some permutation of 1, 1, 1, 2
(2) h, i, j, k is some permutation of 1, 1, 2, 2
(3) h, i, j, k is some permutation of 1, 2, 2, 2

Let us call the sum of the terms of each of these types α, β, and γ, respectively. We therefore have

$$\alpha = \sum_{r,s} \sum_{h,i,j,k} \begin{vmatrix} v_{1hir} & v_{1jks} \\ v_{2hir} & v_{2jks} \end{vmatrix} \begin{vmatrix} 1 & a_r \\ 1 & a_s \end{vmatrix} \quad h,i,j,k \text{ a permutation of } 1,1,1,2$$

$$\beta = \sum_{r,s} \sum_{h,i,j,k} \begin{vmatrix} v_{1hir} & v_{1jks} \\ v_{2hir} & v_{2jks} \end{vmatrix} \begin{vmatrix} 1 & a_r \\ 1 & a_s \end{vmatrix} \quad h,i,j,k \text{ a permutation of } 1,1,2,2$$

$$\gamma = \sum_{r,s} \sum_{h,i,j,k} \begin{vmatrix} v_{1hir} & v_{1jks} \\ v_{2hir} & v_{2jks} \end{vmatrix} \begin{vmatrix} 1 & a_r \\ 1 & a_s \end{vmatrix} \quad h,i,j,k \text{ a permutation of } 1,2,2,2$$

Using these symbols,

$$|VA| = \alpha + \beta + \gamma$$

By (3.22),

$$\text{2-positive determinant} = |\bar{A}'|(\alpha + \beta + \gamma) \qquad (3.23)$$

Now let us choose matrices of order 2 from our initial array in the following manner. The first column of each new matrix will be the first column of one of the squares in the array. The second column of each new matrix will be the second column of one of the squares in the array. The two columns will be chosen such that, in the determinant of the new matrix, each term of the expansion of this determinant will have in its subscripts the number 1 appearing three times. An example of such a determinant is

$$\begin{vmatrix} p_{.1..} & p_{...1} \\ p_{11..} & p_{1..1} \end{vmatrix} = p_{.1..}\,p_{1..1} - p_{...1}\,p_{11..}$$

Notice on the right hand side that the number 1 appears three times in the subscripts of each term. The left hand column of this determinant is the left hand column of square (2), and the right hand column is the right hand column of square (1). There are three other determinants which can be formed in this manner, namely,

$$\begin{vmatrix} 1 & p_{.1.1} \\ p_{1...} & p_{11.1} \end{vmatrix} \quad \begin{vmatrix} 1 & p_{..11} \\ p_{1...} & p_{1.11} \end{vmatrix} \quad \text{and} \quad \begin{vmatrix} p_{..1.} & p_{...1} \\ p_{1.1.} & p_{1..1} \end{vmatrix}$$

We shall call determinants of this type "3-positive" determinants.

Putting these determinants in terms of the latent structure, we get the following equations:

$$\begin{vmatrix} 1 & p_{.1.1} \\ p_{1...} & p_{11.1} \end{vmatrix} = |\bar{A}'| \left| V \begin{pmatrix} 1 & a_1 \cdot a_1 \\ 1 & a_1 \cdot a_2 \\ 1 & a_1 \cdot a_1 \\ 1 & a_1 \cdot a_2 \\ 1 & a_2 \cdot a_1 \\ 1 & a_2 \cdot a_2 \\ 1 & a_2 \cdot a_1 \\ 1 & a_2 \cdot a_2 \end{pmatrix} \right| \tag{3.24}$$

$$\begin{vmatrix} p_{.1..} & p_{...1} \\ p_{11..} & p_{1..1} \end{vmatrix} = |\bar{A}'| \left| V \begin{pmatrix} a_1 & a_1 \\ a_1 & a_2 \\ a_1 & a_1 \\ a_1 & a_2 \\ a_2 & a_1 \\ a_2 & a_2 \\ a_2 & a_1 \\ a_2 & a_2 \end{pmatrix} \right| \tag{3.25}$$

$$\begin{vmatrix} 1 & p_{..11} \\ p_{1...} & p_{1.11} \end{vmatrix} = |\bar{A}'| \left| V \begin{pmatrix} 1 & a_1 \cdot a_1 \\ 1 & a_1 \cdot a_2 \\ 1 & a_2 \cdot a_1 \\ 1 & a_2 \cdot a_2 \\ 1 & a_1 \cdot a_1 \\ 1 & a_1 \cdot a_2 \\ 1 & a_2 \cdot a_1 \\ 1 & a_2 \cdot a_2 \end{pmatrix} \right| \tag{3.26}$$

$$\begin{vmatrix} p_{..1.} & p_{...1} \\ p_{1.1.} & p_{1..1} \end{vmatrix} = |\bar{A}'| \left| V \begin{pmatrix} a_1 & a_1 \\ a_1 & a_2 \\ a_2 & a_1 \\ a_2 & a_2 \\ a_1 & a_1 \\ a_1 & a_2 \\ a_2 & a_1 \\ a_2 & a_2 \end{pmatrix} \right| \tag{3.27}$$

Let us be clear how these equations were derived. The right hand column of the second rectangular matrix (in a) affects the right hand column of the determinant of manifest data, and similarly for the left hand column. Now the manifest data determinants of (3.24) and (3.26) have left hand columns which are the same as that of the 2-positive determinant, hence the left hand columns of the second rectangular matrices in these equations are the same as the left hand column of A. Similarly, the right hand columns of the manifest data in (3.25) and (3.27) are the same as that of the 2-positive determinant, so the right hand columns of the second rectangular matrices are the same as the right hand column of A.

Now the first and fourth subscripts are the same for all determinants, and we do not need to worry about them, since $|\bar{A}'|$ takes care of the first subscript and A takes care of the fourth. Recall that the V matrix was arranged so that its columns are in lexical order with respect to the second and third subscripts. Since a given row of A corresponds to the same column of V, any probabilities which are put into A which refer to the second or third interviews will have to take account of that lexical order. In (3.24) the right hand column of manifest data is the right hand column of the 2-positive determinant stratified by time 2. The lexical order means that the first four columns of V have the subscript 1 in the second place, and the last four columns have the subscript 2 in the second place. Hence the first four rows of the right hand column of A must be multiplied by a_1, and the last four rows of the right hand column of A by a_2. Comparison of (3.24) with (3.21) will show that this is in fact what has been done.

In (3.25) the first column of manifest data is the first column of the 2-positive determinant stratified by time 2, while the second column is the same as that of the 2-positive determinant. Hence the first column of A is multiplied, the first four rows by a_1, the last four rows by a_2.

(3.26) has as its first column the first column of the 2-positive determinant, while the second column is stratified by time 3. Hence the first column of A is unaffected, while the second column has its rows multiplied by a_1 or a_2. The lexical order of second and third subscripts means that for time 3, the first two rows multiply columns of V which have the subscript 1 in the third place, the second two rows multiply columns of V which have the subscript 2 in the third place, the third two rows refer to the subscript 1 in the third place, and the fourth two rows refer to the subscript 2 in the third place. Hence the first two rows of the second column of A are multiplied by a_1, the second two rows by a_2, etc. In (3.27) the same procedure is followed for the first column of A instead of the second, since the manifest determinant in (3.27) has as its first column the first column of the 2-positive determinant stratified by time 3, the second column being the same as that of the 2-positive determinant.

Since V is the same in all four equations, it is evident that in all four expansions of the determinant which is the product of V and a modified A matrix, there will be terms involving the same minors of V. Hence if we add the four 3-positive determinants, the sum will be the product of $|\bar{A}'|$ and a sum of terms which can be collected in sets of four with respect to a minor of V. These collected terms will be of the following type:

$$\begin{vmatrix} v_{1hir} & v_{1jks} \\ v_{2hir} & v_{2jks} \end{vmatrix} \left\{ \begin{vmatrix} 1 & a_h \cdot a_r \\ 1 & a_j \cdot a_s \end{vmatrix} + \begin{vmatrix} a_h & a_r \\ a_j & a_s \end{vmatrix} + \begin{vmatrix} 1 & a_i \cdot a_r \\ 1 & a_k \cdot a_s \end{vmatrix} + \begin{vmatrix} a_i & a_r \\ a_k & a_s \end{vmatrix} \right\}$$

The sum of the first two bracketed determinants, as was shown before (using (3.12) and (3.13)), is equal to

$$(a_h + a_j) \begin{vmatrix} 1 & a_r \\ 1 & a_s \end{vmatrix}$$

Similarly, the sum of the last two determinants is

$$(a_i + a_k) \begin{vmatrix} 1 & a_r \\ 1 & a_s \end{vmatrix}$$

Hence the sum of all four bracketed determinants is

$$(a_h + a_i + a_j + a_k) \begin{vmatrix} 1 & a_r \\ 1 & a_s \end{vmatrix}$$

Therefore the sum of the 3-positive determinants is

$$|\bar{A}'| \sum_{r,s} \sum_{h,i,j,k} (a_h + a_i + a_j + a_k) \begin{vmatrix} v_{1hir} & v_{1jks} \\ v_{2hir} & v_{2jks} \end{vmatrix} \begin{vmatrix} 1 & a_r \\ 1 & a_s \end{vmatrix}$$

Now consider the subscripts h, i, j, and k. When these are all either 1 or 2, the determinant in v vanishes, as was shown above, and therefore the whole product vanishes. When h, i, j, and k are some permutation of 1, 1, 1, and 2,

$$\begin{aligned} a_h + a_i + a_j + a_k &= a_1 + a_1 + a_1 + a_2 \\ &= 3a_1 + a_2 \end{aligned}$$

Hence the sum of terms of this type is

$$|\bar{A}'|\,(3a_1+a_2)\sum_{r,s}\sum_{h,i,j,k}\begin{vmatrix} v_{1hir} & v_{1jks} \\ v_{2hir} & v_{2jks}\end{vmatrix}\begin{vmatrix} 1 & a_r \\ 1 & a_s\end{vmatrix} \quad h,i,j,k \text{ some permutation of } 1,1,1,2$$

By the notation introduced above, this is

$$|\bar{A}'|\,(3a_1+a_2)\alpha$$

When h, i, j, and k are some permutation of 1, 1, 2, and 2,

$$a_h + a_i + a_j + a_k = 2a_1 + 2a_2$$

Hence the sum of terms of this type is equal to

$$|\bar{A}'|\,(2a_1+2a_2)\beta$$

Finally, when h, i, j, and k are some permutation of 1, 2, 2, and 2,

$$a_h + a_i + a_j + a_k = a_1 + 3a_2$$

so that the sum of terms of this type is equal to

$$|\bar{A}'|\,(a_1+3a_2)\gamma$$

Thus the sum of all the 3-positive determinants is as follows:

Σ 3-positive determinants =

$$|\bar{A}'|\,[(3a_1+a_2)\alpha + (2a_1+2a_2)\beta + (a_1+3a_2)\gamma] \tag{3.28}$$

If we divide this by (3.17) we get

$$\frac{\Sigma\ \text{3-positive determinants}}{\text{2-positive determinant}} = \frac{(3a_1+a_2)\alpha + (2a_1+2a_2)\beta + (a_1+3a_2)\gamma}{\alpha+\beta+\gamma} \tag{3.29}$$

But notice that we can take (a_1+a_2) out of the numerator of the right hand side, getting

$$(a_1+a_2)(\alpha+\beta+\gamma) + 2a_1\alpha + (a_1+a_2)\beta + 2a_2\gamma$$

Thus we can express (3.29) as

$$\frac{\Sigma\ \text{3-positive determinants}}{\text{2-positive determinant}} = a_1 + a_2 + \frac{2a_1\alpha + (a_1+a_2)\beta + 2a_2\gamma}{\alpha+\beta+\gamma} \tag{3.30}$$

We now define the 4-positive determinants as follows. The first column of each is the first column of one square of the initial array of manifest data. The second column of each is the second column of one square of the initial array. The columns are chosen so that in the expansion of every 4-positive determinant each term will have the subscript 1 appearing four times. There are six determinants which can be formed in this manner, of which the following are examples:

$$\begin{vmatrix} 1 & p_{.111} \\ p_{1...} & p_{1111} \end{vmatrix} \quad \text{and} \quad \begin{vmatrix} p_{..1.} & p_{.1.1} \\ p_{1.1.} & p_{11.1} \end{vmatrix}$$

Each of these determinants is expressible as the product of $|\bar{A}'|$ and the determinant of the product of two rectangular matrices, the first of which is V, and the second of which is A with certain rows and columns multiplied by a_1, a_2 or powers or products of a_1 and a_2. The expansions of these determinants of the product of matrices will all have terms involving the same minors of V, hence the sum of all the 4-positive determinants can be expressed as the product of $|\bar{A}'|$ and a sum of terms, each of which is a product of some minor of V and the sum of six other determinants. These products will be of the following type:

$$\begin{vmatrix} v_{1hir} & v_{1jks} \\ v_{2hir} & v_{2jks} \end{vmatrix} \left\{ \begin{vmatrix} a_h & a_i a_r \\ a_j & a_k a_s \end{vmatrix} + \begin{vmatrix} a_i & a_h a_r \\ a_k & a_j a_s \end{vmatrix} + \begin{vmatrix} 1 & a_h a_i a_r \\ 1 & a_j a_k a_s \end{vmatrix} \right.$$

$$\left. + \begin{vmatrix} a_h a_i & a_r \\ a_j a_k & a_s \end{vmatrix} + \begin{vmatrix} a_h & a_h a_r \\ a_j & a_j a_s \end{vmatrix} + \begin{vmatrix} a_i & a_i a_r \\ a_k & a_k a_s \end{vmatrix} \right\}$$

The sum of the first two bracketed determinants is

$$(a_h a_k + a_i a_j) \begin{vmatrix} 1 & a_r \\ 1 & a_s \end{vmatrix}$$

The sum of the second two bracketed determinants is

$$(a_j a_k + a_h a_i) \begin{vmatrix} 1 & a_r \\ 1 & a_s \end{vmatrix}$$

The sum of the third two bracketed determinants is

$$(a_h a_j + a_i a_k) \begin{vmatrix} 1 & a_r \\ 1 & a_s \end{vmatrix}$$

Therefore the sum of the 4-positive determinants is

$$|\bar{A}'| \sum_{r,s} \sum_{h,i,j,k} (a_h a_i + a_h a_j + a_h a_k + a_i a_j + a_i a_k + a_j a_k)$$

$$\cdot \begin{vmatrix} v_{1hir} & v_{1jks} \\ v_{2hir} & v_{2jks} \end{vmatrix} \begin{vmatrix} 1 & a_r \\ 1 & a_s \end{vmatrix}$$

When h, i, j, and k are a permutation of 1, 1, 1, and 2, there are three ways in which the 2 can pair with a 1 and three ways in which a 1 can pair with another 1. Therefore

$$a_h a_i + a_h a_j + a_h a_k + a_i a_j + a_i a_k + a_j a_k = (3a_1^2 + 3a_1 a_2)$$

When h, i, j, and k are some permutation of 1, 1, 2, and 2, there is one way in which a 1 can pair with a 1, there are four ways in which a 1 can pair with a 2, and there is one way in which a 2 can pair with a 2. Hence the expression in parentheses above equals

$$a_1^2 + 4a_1 a_2 + a_2^2$$

When h, i, j, and k are some permutation of 1, 2, 2, and 2, there are three ways in which a 1 can pair with a 2 and three ways in which a 2 can pair with a 2. Then the expression in parentheses above is

$$3a_1 a_2 + 3a_2^2$$

Therefore the sum of all 4-positive determinants is as follows:

$$\Sigma \text{ 4-positive determinants} =$$

$$|\bar{A}'| \, [(3a_1^2 + 3a_1 a_2)\alpha + (a_1^2 + 4a_1 a_2 + a_2^2)\beta + (3a_1 a_2 + 3a_2^2)\gamma]$$

Dividing by (3.23) gives

$$\frac{\Sigma \text{ 4-positive determinants}}{\text{2-positive determinant}} =$$

$$\frac{(3a_1^2 + 3a_1 a_2)\alpha + (a_1^2 + 4a_1 a_2 + a_2^2)\beta + (3a_1 a_2 + 3a_2^2)\gamma}{\alpha + \beta + \gamma} \qquad (3.31)$$

The 5-positive determinants are defined and expanded and summed in similar fashion. There are four such determinants, and their sum is equal to

$$|\bar{A}'| \sum_{r,s} \cdot \sum_{h,i,j,k} (a_h a_i a_j + a_h a_i a_k + a_h a_j a_k + a_i a_j a_k)$$

$$\cdot \begin{vmatrix} v_{1hir} & v_{1jks} \\ v_{2hir} & v_{2jks} \end{vmatrix} \begin{vmatrix} 1 & a_r \\ 1 & a_s \end{vmatrix}$$

When h, i, j, k is a permutation of 1, 1, 1, 2, there is one triplet with subscripts all 1, and there are three triplets with two subscripts 1 and one subscript 2, which makes the expression in parentheses

$$(a_1^3 + 3a_1^2 a_2)$$

When h, i, j, k is a permutation of 1, 1, 2, 2, the expression in parentheses becomes

$$(2a_1^2 a_2 + 2a_1 a_2^2)$$

and when h, i, j, k is a permutation of 1, 2, 2, 2, the expression in parentheses becomes

$$(3a_1 a_2^2 + a_2^3)$$

Hence the sum of 5-positive determinants is

$$|\bar{A}'| [(a_1^3 + 3a_1^2 a_2)\alpha + (2a_1^2 a_2 + 2a_1 a_2^2)\beta + (3a_1 a_2^2 + a_2^3)\gamma]$$

Dividing this sum by the 2-positive determinant, (3.23), gives

$$\frac{\Sigma \text{ 5-positive determinants}}{\text{2-positive determinant}} =$$

$$\frac{(a_1^3 + 3a_1^2 a_2)\alpha + (2a_1^2 a_2 + 2a_1 a_2^2)\beta + (3a_1 a_2^2 + a_2^3)\gamma}{\alpha + \beta + \gamma} \tag{3.32}$$

Evaluating the 6-positive determinant (there is only one) in the same manner gives as its sum

$$|\bar{A}'| [a_1^3 a_2 \alpha + a_1^2 a_2^2 \beta + a_1 a_2^3 \gamma]$$

$a_1 a_2$ can be factored out, giving

$$|\bar{A}'| a_1 a_2 (a_1^2 \alpha + a_1 a_2 \beta + a_2^2 \gamma)$$

Dividing by (3.23),

$$\frac{\text{6-positive determinant}}{\text{2-positive determinant}} = a_1 a_2 \left[\frac{a_1^2\alpha + a_1 a_2\beta + a_2^2\gamma}{\alpha + \beta + \gamma}\right] \tag{3.33}$$

Now let

$$x + y = \frac{2a_1\alpha + (a_1 + a_2)\beta + 2a_2\gamma}{\alpha + \beta + \gamma}$$

and

$$xy = \frac{a_1^2\beta + a_1 a_2\beta + a_2^2\gamma}{\alpha + \beta + \gamma}$$

We shall prove that a_1, a_2, x and y are the roots of the equation

$$z^4 - \frac{\Sigma\ \text{3-positive determinants}}{\text{2-positive determinant}} z^3$$

$$+ \frac{\Sigma\ \text{4-positive determinants}}{\text{2-positive determinant}} z^2$$

$$- \frac{\Sigma\ \text{5-positive determinants}}{\text{2-positive determinant}} z$$

$$+ \frac{\text{6-positive determinants}}{\text{2-positive determinant}} = 0 \tag{3.34}$$

Consider the elementary symmetric functions of a_1, a_2, x, and y. We note by inspection of (3.30) and by definition of $(x + y)$ that

$$a_1 + a_2 + x + y = a_1 + a_2 + \left[\frac{2a_1\alpha + (a_1 + a_2)\beta + 2a_2\gamma}{\alpha + \beta + \gamma}\right]$$

$$= \frac{\Sigma\ \text{3-positive determinants}}{\text{2-positive determinant}}$$

The next elementary symmetric function is

$$a_1a_2 + a_1x + a_1y + a_2x + a_2y + xy = a_1a_2 + a_1(x+y) + a_2(x+y) + xy$$

$$= a_1a_2 + (a_1 + a_2)(x+y) + xy$$

$$= \frac{a_1a_2(\alpha+\beta+\gamma) + (a_1+a_2)[2a_1\alpha + (a_1+a_2)\beta + 2a_2\gamma]}{\alpha + \beta + \gamma}$$

$$+ \frac{a_1^2\alpha + a_1a_2\beta + a_2^2\gamma}{\alpha + \beta + \gamma}$$

$$= \frac{(3a_1^2 + 3a_1a_2)\alpha + (a_1^2 + 4a_1a_2 + a_2^2)\beta + (3a_1a_2 + 3a_2^2)\gamma}{\alpha + \beta + \gamma}$$

$$= \frac{\Sigma \text{ 4-positive determinants}}{\text{2-positive determinant}} \qquad \text{by (3.31)}$$

The next elementary symmetric function is

$$a_1a_2x + a_1a_2y + a_1xy + a_2xy = a_1a_2(x+y) + (a_1+a_2)xy$$

$$= \frac{a_1a_2[2a_1\alpha + (a_1+a_2)\beta + 2a_2\gamma]}{\alpha + \beta + \gamma} + \frac{(a_1+a_2)(a_1^2\alpha + a_1a_2\beta + a_2^2\gamma)}{\alpha + \beta + \gamma}$$

$$= \frac{(a_1^3 + 3a_1^2a_2)\alpha + (2a_1^2a_2 + 2a_1a_2^2)\beta + (3a_1a_2^2 + a_2^3)\gamma}{\alpha + \beta + \gamma}$$

$$= \frac{\Sigma \text{ 5-positive determinants}}{\text{2-positive determinant}} \qquad \text{by (3.32)}$$

The next elementary symmetric function is

$$a_1a_2xy = a_1a_2\left(\frac{a_1^2\alpha + a_1a_2\beta + a_2^2\gamma}{\alpha + \beta + \gamma}\right)$$

$$= \frac{\text{6-positive determinant}}{\text{2-positive determinant}} \qquad \text{by (3.33)}$$

Since the elementary symmetric functions, properly signed and placed, are the coefficients of (3.34), a_1, a_2, x, and y are the roots of the equation (3.34). [1] In order to find a_1 and a_2 it is merely necessary to solve this quartic approximately by Horner's method or some other method, keeping in mind that the desired roots will usually be in the neighborhood of 0.10 and 0.90, as an approximation with which to start the solution.

Unfortunately, two other roots are also present, and the question arises which roots are a_1 and a_2. It may be that there are only two real roots between 0 and 1, in which case these are the correct roots. There may be two double roots of the proper sort, which poses no problem. If there are four real roots between 0 and 1, the best that can be offered at this writing is to choose the smallest

[1] An alternative and simpler proof that a_1 and a_2 are roots of (3.34) is given in Appendix A. The present approach stresses the method of formation of the coefficients and is oriented towards generality.

root for a_2 and the largest root for a_1. The selection will have to be a matter of judgment.

Once a_1 and a_2 are determined, it is easy to find the true proportions from the following equations: [1]

$$v_{1...} = \frac{p_{1...} - a_2}{a_1 - a_2}, \text{ etc.}$$

$$v_{11..} = \frac{p_{11..} - a_2(p_{1...} + p_{.1..}) + a_2^2}{(a_1 - a_2)^2}, \text{ etc.}$$

$$v_{111.} = \frac{p_{111.} - a_2(p_{11..} + p_{1.1.} + p_{.11.})}{(a_1 - a_2)^3}$$

$$+ \frac{a_2^2(p_{1...} + p_{.1..} + p_{..1.}) - a_2^3}{(a_1 - a_2)^3}, \text{ etc.}$$

$$v_{1111} = \frac{v_{111.}\, v_{.111}}{v_{.11.}} \qquad (3.35)$$

From this set can be determined all sixteen patterns of membership in the latent classes over the entire time, if they are required.

Example of the Two Class, Four Interview Model

We shall apply the present model to the data on "winner expectations" from the 1940 voting study in Erie County, Ohio, which we used for the previous model. There we saw that the assumption, that when the latent class at time 2 was fixed the latent position at time 3 was independent of the latent position at time 1, did not seem to be warranted. Now we are assuming that when the latent position at time 3 and time 2 is fixed, the latent position at time 4 is independent of the latent position at time 1. Since we have six interviews, there are three successive sets of four interviews we can use to test the assumptions of this model.

We may first comment on a possible objection to the use of a model which involves computations based on quadruple occurrences, or fourth order frequencies. It may be thought that such frequencies will be too small to be useful, but such is not the case.

[1] These equations are derived in Appendix A.

The situation is unlike that where latent structure analysis is applied to a number of different items, because an item is usually much more closely related to itself at another time than it is to most other items at the same time. In the three sets of four interviews we are using now, for example, p_{1111} is successively 0.21, 0.23, and 0.23. As a matter of fact, the probability of a "Democratic" response on all six interviews is 0.16. The use of higher order frequencies, therefore, is not a matter for great concern.

When the proper coefficients of the "winner expectations" data are computed for the successive sequences of interviews, the following equations result:

May—June—July—Aug. $x^4 - 1.8701x^3 + 0.9813x^2 - 0.1050x + 0.01293 = 0$

June—July—Aug.—Sept. $x^4 - 1.8037x^3 + 0.9103x^2 - 0.03921x + 0.003725 = 0$

July—Aug.—Sept.—Oct. $x^4 - 2.0212x^3 + 1.1888x^2 - 0.1538x + 0.01060 = 0$

It turns out that none of these equations has any real roots. When they are graphed, however, each equation has all three bends between the values 0 and 1 of the argument. All three curves reach a minimum at a low positive value of x, a maximum at about $x = 0.5$, and another minimum between $x = 0.5$ and $x = 1$. In the absence of a statistical theory, we do not know how close these minima must be to the x-axis for the discrepancy to be considered a chance error. The value of the function is less than 0.01 at the first minimum for each equation. For the minimum which is between $x = 0.5$ and $x = 1$, the value of the function is less than 0.01 for the May—June—July—August sequence, less than 0.02 for the July—August—September—October sequence, but greater than 0.04 for the June—July—August—September sequence. Since the maxima for all three curves are around 0.04 or 0.05, it seems reasonable to accept the minima below 0.01, and possibly the one between 0.01 and 0.02, as chance discrepancies from the x-axis. On this basis we have to reject the model for the June—July—August—September sequence. For the other sequences, we take the two minima as the probabilities of positive response. This gives the following result:

Latent Class	*Latent probabilities*	
	May—June—July—Aug.	*July—Aug.—Sept.—Oct.*
Democratic	0.91	0.91
Not Democratic	0.94	0.92

The two sequences give very similar latent probabilities but the failure of the middle sequence to conform partially vitiates the confirmation. Notice also that the probabilities obtained here are lower than those obtained for the same data using the previous model, where they averaged about 0.95, omitting the June—July—August sequence where the model obviously did not apply. The reason for this drop is probably not that respondents reverted over the four waves. If the latent position at time 3 is not independent of the latent position at time 1 when the position at time 2 is fixed, the main effect of such a dependence would not be real reversion, which would spuriously reduce the latent probability, but rather a tendency to adhere to the original position throughout the sequence, which would spuriously raise the probability. The introduction of the present model would reduce that spurious element.

This observation gives us a basis for a testing procedure for the independence assumptions, if we have a large number of interviews available. If the latent position at a given time is independent of the latent position m intervals previously, it will also be independent of the latent position $m + 1$ intervals previously. In that case the latent probabilities obtained on these two different assumptions of independence should be the same. The proper procedure to test such assumptions would be to take successively longer sequences, in each case assuming that the terminal latent position is independent of the initial latent position, and compute the probabilities. As long as the results differ substantially for successive computations, the assumption must be rejected for the shorter sequence. When the probabilities obtained from a sequence of m interviews are the same as those for a sequence of $m - 1$ interviews, the assumption may be held valid (if the rest of the model is valid).

N Interviews, Two Latent Classes, Terminal Latent Position Independent of Initial Latent Position, Controlling Intermediate Latent Positions

The solution of the four interview case was stated in such a way as to be immediately generalized, by analogy, to the case of any larger number of interviews. Consider, for example, the five interview case. Let us change the notation for the manifest data as follows. Since we use only the proportions giving the positive response, let the subscripts of p refer by number to the time, so that p_{145}, e.g., will mean the proportion giving the positive response on the first, fourth, and fifth interviews of a sequence. Then arrange the manifest data in the following way:

(1)		(2)		(3)		(4)	
1	p_5	p_2	p_{25}	p_3	p_{35}	p_4	p_{45}
p_1	p_{15}	p_{12}	p_{125}	p_{13}	p_{135}	p_{14}	p_{145}

(5)		(6)		(7)		(8)	
p_{23}	p_{235}	p_{24}	p_{245}	p_{34}	p_{345}	p_{234}	p_{2345}
p_{123}	p_{1235}	p_{124}	p_{1245}	p_{134}	p_{1345}	p_{1234}	p_{12345}

The numbers in parentheses are merely put in to identify the squares. The right hand column of each square is stratified by the terminal interview, number 5. The bottom row of each square is stratified by the initial interview, number 1. The first square, which is the 2-positive matrix, contains only the stratifying elements. The next three squares refer to the marginal frequencies for each time between the initial and terminal interviews, properly stratified in the right hand column and bottom row. The next three squares, (5), (6), and (7), refer to the joint frequencies for all pairs of interviews between the first and last, properly stratified in the right hand column and bottom row. The final square (8) refers to the triple occurrence frequency for all three waves except the first and last, properly stratified. The number of squares of each type is simply the proper binomial coefficient for the power equal to the number of interviews between the first and last.

The determinant of (1) is the 2-positive, by which all other coefficients are divided. Now every determinant we use will have as its left hand column the left hand column of one of the eight squares, and as its right hand column the right hand column of one of the eight squares. In order to form the 3-positive determinants, we take every pair of right and left columns which will form a determinant whose evaluated terms will consist of the product of two factors p, such that the total number of subscripts of both p's in the product will be three. In the present case, such determinants can only be formed by using the left hand column of (1) with the right hand column of (2), (3), and (4), successively, and by using the right hand column of (1) with the left hand column of (2), (3), and (4), successively. There will thus be six 3-positive determinants.

To form the 4-positives, proceed in the same manner. (1) can be combined with (5), (6), and (7) to form six such determinants. (2), (3), and (4) are 4-positives taken by themselves and combined with each other. Since there are three left hand columns and three right hand columns which can be combined in any way, there are nine 4-positive determinants which can be formed from (2), (3), and (4), making a total of fifteen 4-positive determinants which can be formed. Continuing in this manner, we get twenty 5-positive determinants, fifteen 6-positives, six 7-positives, and finally one 8-positive, which is the determinant of (8). Since we have seven sets of determinants, running from 2-positive through 8-positive, it is evident that we will have an equation of the sixth degree. Notice that the number of determinants of each type is simply the binomial coefficient of the expansion to the sixth power.

The resulting equation takes the form

$$\begin{aligned} x^6 &- \frac{\Sigma\ 6\ \text{3-positive determinants}}{\text{2-positive determinant}}\, x^5 \\ &+ \frac{\Sigma\ 15\ \text{4-positive determinants}}{\text{2-positive determinant}}\, x^4 \\ &- \frac{\Sigma\ 20\ \text{5-positive determinants}}{\text{2-positive determinant}}\, x^3 \\ &+ \frac{\Sigma\ 15\ \text{6-positive determinants}}{\text{2-positive determinant}}\, x^2 \\ &- \frac{\Sigma\ 6\ \text{7-positive determinants}}{\text{2-positive determinant}}\, x \\ &+ \frac{\text{8-positive determinant}}{\text{2-positive determinant}} = 0 \end{aligned} \tag{3.36}$$

a_1 and a_2 are roots of this equation; there are also four spurious roots. If the model holds perfectly, these roots would be two equal to a_1 and two equal to a_2. Again, judgment must be used in selecting the roots. However, it may often be unnecessary to bother with the spurious roots, since they may fall outside the reasonable range of latent probabilities. The probabilities computed for shorter sequences of the same data provide useful approximations with which to begin Horner's method, which is not too difficult to apply when such approximations are available.

Once a_1 and a_2 are known, the latent proportions can be found from the fundamental equations, as in previous cases. From these the probabilities of true change can be found, if desired.

Analogues for Continuous Variables to Models of this Chapter

The assumption that true (or latent) change occurs in only one direction does not have much meaning for continuous variables, since, as noted in Chapter 2, if everyone shifts in the same direction and maintains his relative position in the distribution, the correlations are unaffected. We go on directly to the case where it is assumed that changes of true scores occur in any direction.

The first case for which solutions can be obtained is that in which three waves (or time points) are available, and in which it is assumed that the true scores at time 3 are independent of those at time 1 when the true scores at time 2 are controlled. We assume that the variable is normally distributed at each time. Then, using the notation of Chapter 2, the assumption means that

$$r_{T_1T_3 \cdot T_2} = 0 \tag{3.37}$$

where the standard notation for partial correlation is used, in which the correlated variables are shown in subscripts to the left of the dot and the control variables are shown in subscripts to the right of the dot.

We can now make use of the standard formula for partial correlation.

$$r_{T_1T_3 \cdot T_2} = \frac{r_{T_1T_3} - r_{T_1T_2}\ r_{T_2T_3}}{\sqrt{1-r^2_{T_1T_2}}\sqrt{1-r^2_{T_2T_3}}} \tag{3.38}$$

Since this is equal to zero by (3.37), the numerator must equal zero, and

$$r_{T_1T_3} = r_{T_1T_2}\ r_{T_2T_3} \tag{3.39}$$

Written out, this is

$$\frac{s_{T_1T_3}}{s_{T_1}s_{T_3}} = \frac{s_{T_1T_2}}{s_{T_1}s_{T_2}} \cdot \frac{s_{T_2T_3}}{s_{T_2}s_{T_3}} \tag{3.40}$$

Multiplying through by $s_{T_1} s_{T_3}$, and rearranging,

$$s^2_{T_2} = \frac{s_{T_1T_2} \cdot s_{T_2T_3}}{s_{T_1T_3}} \tag{3.41}$$

i.e., the true variance at time 2 equals the product of the true covariances for times 1 and 2 and times 2 and 3 divided by the true covariance for times 1 and 3.

Since the errors are uncorrelated with anything, the observed covariances are equal to the corresponding true covariances, as implied by (2.61). By substituting the observed covariance in (3.41), the true variance at time 2 can be computed, as follows:

$$s^2_{T_2} = \frac{s_{X_1X_2} \cdot s_{X_2X_3}}{s_{X_1X_3}} \tag{3.42}$$

By (2.57),

$$r_{r_2} = s^2_{T_2}/s^2_{X_2}$$

for time 2. r_{r2} is the reliability at time 2, where it must be kept in mind that we are using the word unreliability not in its usual sense as referring merely to error of measurement, but as including any random component in the attitude or behavior. By assumption, the reliability is the same at time 1, time 2, and time 3.

Since the observed variance at time 2 can be computed, we have

$$r_{r_2} = \frac{s_{X_1X_2} \cdot s_{X_2X_3}}{s^2_{X_2} \cdot s_{X_2X_3}} \tag{3.43}$$

This is the solution for this particular model. Let us consider a few further implications of the model. The first order partial correlation of the observed scores will be

$$r_{X_1X_3 \cdot X_2} = \frac{r_{X_1X_3} - r_{X_1X_2} r_{X_2X_3}}{\sqrt{1 - r^2_{X_1X_2}} \sqrt{1 - r^2_{X_2X_3}}} \tag{3.44}$$

Consider only the numerator of this expression on the right. Written out, it is

$$\frac{s_{X_1X_3}}{s_{X_1} s_{X_3}} - \frac{s_{X_1X_2}}{s_{X_1} s_{X_2}} \cdot \frac{s_{X_2X_3}}{s_{X_2} s_{X_3}} \tag{3.45}$$

This is equal to

$$\frac{s_{X_2}^2 s_{X_1X_3} - s_{X_1X_2} s_{X_2X_3}}{s_{X_1} s_{X_2}^2 s_{X_3}} \tag{3.46}$$

Consider only the numerator of (3.46). The observed covariances $s_{X_iX_j}$ are equal to the corresponding true covariances $s_{T_iT_j}$. By (2.56), the observed variance at time 2 is

$$s_{X_2}^2 = s_{T_2}^2 + s_{E_2}^2 \tag{3.47}$$

Hence the numerator of (3.46) becomes

$$\begin{aligned}(s_{T_2}^2 + s_{E_2}^2)s_{T_1T_3} - s_{T_1T_2} s_{T_2T_3} \\ = (s_{T_2}^2 s_{T_1T_3} - s_{T_1T_2} s_{T_2T_3}) + s_{E_2}^2 s_{T_1T_3}^2\end{aligned} \tag{3.48}$$

But by (3.41) the term in parentheses on the right hand side of (3.48) is equal to zero. Hence the numerator of (3.46) is greater than zero, and consequently the whole term is greater than zero (since variances and standard deviations are positive). This in turn implies that (3.44) is greater than zero, or

$$r_{X_1X_3 \cdot X_2} > 0 \tag{3.49}$$

In other words, the partial correlation of the observed scores will be greater than zero, under the assumptions of this model, although the partial correlation of the true scores is equal to zero.

If more than three time points are available, the model imposes certain conditions on the data. Considering first the true correlations, we have

$$r_{T_1T_3} = r_{T_1T_2} r_{T_2T_3}$$

by (3.39). But under the assumptions, the partial correlation of true scores at times 1 and 4 controlling the true scores at time 2 will also be zero, so that

$$r_{T_1T_4} = r_{T_1T_2} r_{T_2T_4}$$

From these two equations, solving each for $r_{T_1T_2}$, we get

$$r_{T_1T_2} = \frac{r_{T_1T_3}}{r_{T_2T_3}} = \frac{r_{T_1T_4}}{r_{T_2T_4}}$$

or

$$\begin{vmatrix} r_{T_1 T_3} & r_{T_1 T_4} \\ r_{T_2 T_3} & r_{T_2 T_4} \end{vmatrix} = 0 \tag{3.50}$$

Generalizing this, we can say that the matrix of true correlations has unit rank above the principal diagonal (see below), under the assumptions of this model. What about the matrix of observed correlations? First let us note again that the assumptions of the models of this chapter imply that the reliability is the same at each time. This is equivalent to the assumption in the discrete case that the latent probabilities do not change.

Now multiply both sides of (3.40) by the following term:

$$\frac{s_{T_1} s_{T_2}}{s_{X_1} s_{X_2}} \cdot \frac{s_{T_2} s_{T_3}}{s_{X_2} s_{X_3}}$$

This gives

$$\frac{s_{T_2}^2}{s_{X_2}^2} \cdot \frac{s_{T_1 T_3}}{s_{X_1} s_{X_3}} = \frac{s_{T_1 T_2}}{s_{X_1} s_{X_2}} \cdot \frac{s_{T_2 T_3}}{s_{X_2} s_{X_3}} \tag{3.51}$$

By (2.57)

$$s_{T_2}^2 / s_{X_2}^2 = r_{r_2} = r_r$$

Since the true covariances are equal to the observed covariances, (3.51) is

$$r_r r_{X_1 X_3} = r_{X_1 X_2} r_{X_2 X_3} \tag{3.52}$$

Also,

$$\frac{r_{X_i X_j}}{r_{T_i T_j}} = \frac{s_{X_i X_j} / s_{X_i} s_{X_j}}{s_{T_i T_j} / s_{T_i T_j}}$$

$$= \frac{s_{T_i}}{s_{X_i}} \cdot \frac{s_{T_j}}{s_{X_j}}$$

$$= \sqrt{r_{r_i}} \sqrt{r_{r_j}}$$

$$= r_r \tag{3.53}$$

under the present assumption that all r_r are equal at all times. Thus the fundamental relationship is that

$$r_{X_iX_j} = r_r r_{T_iT_j} \tag{3.54}$$

Substituting such relations in (3.52) gives us (3.39).

Just as (3.52) was derived from (3.39), so the following equation can be derived from the fact that under the assumptions

$$\begin{aligned} r_{T_1T_4} &= r_{T_1T_3} r_{T_3T_4} \\ r_r r_{X_1X_4} &= r_{X_1X_3} r_{X_3X_4} \end{aligned} \tag{3.55}$$

Substituting in this the value of $r_{X_1X_3}$ from (3.52) gives

$$r_r^2 r_{X_1X_4} = r_{X_1X_2} r_{X_2X_3} r_{X_3X_4} \tag{3.56}$$

This can be iterated as far as the matrix extends. Substituting into this value from (3.54) and cancelling out r^3 gives

$$r_{T_1T_4} = r_{T_1T_2} r_{T_2T_3} r_{T_3T_4} \tag{3.57}$$

and so on. Finally, putting observed correlations on the left hand side and true correlations on the right hand side, relations of the following sort are obtained:

$$\begin{aligned} r_{X_iX_j} &= r_r r_{T_iT_j} \\ r_{X_iX_{i+m}} &= r_r r_{T_iT_{i+1}} r_{T_{i+1}T_{i+2}} \cdots r_{T_{i+m-1}T_{i+m}} \end{aligned} \tag{3.58}$$

We can now see what the matrix of observed correlations will look like. Putting in values from (3.58) for the case where four time points are available gives the following correlation matrix over time:

	Time			
Time	1	2	3	4
	(r_r)	$r_r r_{T_1T_2}$	$r_r r_{T_1T_2} r_{T_2T_3}$	$r_r r_{T_1T_2} r_{T_2T_3} r_{T_3T_4}$
	$r_r r_{T_1T_2}$	(r_r)	$r_r r_{T_2T_3}$	$r_r r_{T_2T_3} r_{T_3T_4}$
	$r_r r_{T_1T_2} r_{T_2T_3}$	$r_r r_{T_2T_3}$	(r_r)	$r_r r_{T_3T_4}$
	$r_r r_{T_1T_2} r_{T_2T_3} r_{T_3T_4}$	$r_r r_{T_2T_3} r_{T_3T_4}$	$r_r r_{T_3T_4}$	(r_r)

The inferred reliabilities have been put in parentheses in the principal diagonal. It is easy to see that the matrix has unit rank above the principal diagonal, by looking, say, at the minor of order two in the upper right hand corner of the matrix. It can also be seen that the matrix does not have unit rank across the diagonal. In fact, the matrix looks very much like a cross-product matrix for a latent distance model of latent structure analysis, having also the characteristic "roof-top" appearance in which the correlations are highest near the diagonal and fall off to the northeast and southwest.

It can also be seen that the solution for r_r involves estimating the diagonal so as to maintain unit rank of the matrix. The similarities to factor analysis and latent structure analysis are evident (see Thurstone, 1949, and Lazarsfeld and Henry, 1968).

Since it is desirable to use all the data in obtaining a solution, the proper procedure if more than three panel waves are available would be to test the matrix first for unit rank, then to average the solutions obtained from all possible determinants. It is even easier to sum all products of the type $r_{X_iX_j}r_{X_jX_k}$ and then sum an equivalent number of $r_{X_iX_k}$, and divide the former by the latter.

As an example of the use of the present model, we have chosen data from the class panel (described in Appendix C; also see Wiggins, 1954b) which consists of responses from 71 graduate students in Sociology at Columbia University to the question "Which of the possible grades for this course would you consider a 'good' grade, not in general, but for yourself, in terms of your long-range objectives?" We may call this variable "goal-aspiration." Data were obtained at ten times throughout the school year, at intervals of three weeks. The first four waves will be used in the present example. The matrix of correlations observed is shown in Table 9.

TABLE 9

Correlation matrix, level of aspiration at four times

	1	*2*	*3*	*4*
1	/	0.71	0.67	0.62
2		/	0.85	0.79
3			/	0.90
4				/

$N = 71$

There is one complete determinant of order two, which should be approximately equal to zero if the model is to be used. We have

$$r_{X_1X_3}r_{X_2X_4} = 0.67 \cdot 0.79 = 0.529$$

$$r_{X_1X_4}r_{X_2X_3} = 0.62 \cdot 0.85 = 0.527$$

With a value of 0.002, the determinant is close enough to zero to warrant the assumptions.

There are four equations which may be used for solutions. Combining all the data to get a single solution, we have

$$r_r = \frac{r_{X_1X_2}r_{X_2X_3} + r_{X_1X_2}r_{X_2X_4} + r_{X_1X_3}r_{X_2X_4} + r_{X_2X_3}r_{X_3X_4}}{r_{X_1X_3} + r_{X_1X_4} + r_{X_1X_4} + r_{X_2X_4}}$$

$$= \frac{(0.71)(0.85) + (0.71)(0.79) + (0.67)(0.90) + (0.85)(0.90)}{0.67 + 0.62 + 0.62 + 0.79}$$

$$= 2.533/2.70$$

$$r_r = 0.936$$

This method seems all right, but a little more information can be obtained by computing each estimate of reliability separately. Of the four computing equations, two provide what are essentially estimates of r_{r2} and two provide estimates of r_{r3}. By computing them separately, we can check the assumptions that the reliability does not change. We obtain these results:

Estimates of r_{r_2}	Estimates of r_{r_3}
$\frac{r_{X_1X_2}r_{X_2X_3}}{r_{X_1X_3}} = 0.900$	$\frac{r_{X_1X_3}r_{X_3X_4}}{r_{X_1X_4}} = 0.972$
$\frac{r_{X_1X_2}r_{X_2X_4}}{r_{X_1X_4}} = 0.905$	$\frac{r_{X_2X_3}r_{X_3X_4}}{r_{X_2X_4}} = 0.968$

We see that the two estimates of r_{r2} give about the same result, about 0.90, and the two estimates of r_{r3} give about the same result, about 0.97. This makes us feel that the assumption that the reliability is not changing is false, even though the assumption about the independence of true scores at times 1 and 3 when true scores at time 2 are held constant seems to be warranted.

Actually, for all assumptions to be true, there are further conditions on the data, which we shall not discuss in this paper. We shall, however, discuss later the case of changing reliabilities (or latent probabilities).

Meanwhile, following analogies to the models for the discrete case in this chapter, we go on to consider the case with four waves where, in addition to the general assumption that the reliability remains constant, the only assumption made is that the true score at time 4 is independent of the true score at time 1 when the true scores at times 2 and 3 are held constant. This means that

$$r_{T_1 T_4 \cdot T_2 T_3} = 0 \tag{3.59}$$

Writing out one of the formulas for this second order partial correlation, we have

$$r_{T_1 T_4 \cdot T_2 T_3} = \frac{r_{T_1 T_4 \cdot T_3} - r_{T_1 T_2 \cdot T_3}\, r_{T_2 T_4 \cdot T_3}}{\sqrt{1 - r^2_{T_1 T_2 \cdot T_3}}\sqrt{1 - r^2_{T_2 T_4 \cdot T_3}}} \tag{3.60}$$

Since the second order partial correlation equals zero, the numerator of the expression on the right hand side of (3.60) must be equal to zero, or

$$r_{T_1 T_4 \cdot T_3} = r_{T_1 T_2 \cdot T_3}\, r_{T_2 T_4 \cdot T_3} \tag{3.61}$$

Substituting the formulas for these partial correlations in terms of zero-order correlations, we get

$$\frac{r_{T_1 T_4} - r_{T_1 T_3}\, r_{T_3 T_4}}{\sqrt{1 - r^2_{T_1 T_3}}\sqrt{1 - r^2_{T_3 T_4}}} = \frac{r_{T_1 T_2} - r_{T_1 T_3}\, r_{T_2 T_3}}{\sqrt{1 - r^2_{T_1 T_3}}\sqrt{1 - r^2_{T_2 T_3}}} \cdot \frac{r_{T_2 T_4} - r_{T_2 T_3}\, r_{T_3 T_4}}{\sqrt{1 - r^2_{T_2 T_3}}\sqrt{1 - r^2_{T_3 T_4}}} \tag{3.62}$$

Collecting terms and re-arranging, this becomes

$$r_{T_1 T_4} - r_{T_1 T_2} r_{T_2 T_4} - r_{T_1 T_3} r_{T_3 T_4} + r_{T_1 T_2} r_{T_2 T_3} r_{T_3 T_4} + r_{T_1 T_3} r_{T_2 T_3} r_{T_2 T_4} - r_{T_1 T_4} r^2_{T_2 T_3} = 0$$

The expression on the left hand side is nothing but the following third order determinant of the matrix of true correlations, with unity put into the principal diagonal (of the whole original matrix)

$$\begin{vmatrix} r_{T_1T_2} & r_{T_1T_3} & r_{T_1T_4} \\ 1 & r_{T_1T_3} & r_{T_2T_4} \\ r_{T_2T_3} & 1 & r_{T_3T_4} \end{vmatrix} = 0 \tag{3.63}$$

This determinant will still be equal to zero if every element is multiplied by r_r. From (3.54), we know that an observed correlation equals the corresponding true correlation multiplied by r_r, so that this multiplication gives us

$$\begin{vmatrix} r_{X_1X_2} & r_{X_1X_3} & r_{X_1X_4} \\ r_r & r_{X_2X_3} & r_{X_2X_4} \\ r_{X_2X_3} & r_r & r_{X_3X_4} \end{vmatrix} = 0 \tag{3.64}$$

This is a quadratic equation in r_r which can be solved by ordinary algebra. We have, in fact

$$r_r = (0.5r_{X_1X_4})[r_{X_1X_2}r_{X_2X_4} + r_{X_1X_3}r_{X_3X_4} \pm \\ \pm \sqrt{\begin{matrix}(r_{X_1X_2}r_{X_2X_4}r_{X_1X_3}r_{X_3X_4}) \\ -4r_{X_1X_4}r_{X_2X_3}(r_{X_1X_2}r_{X_3X_4} + r_{X_1X_3}r_{X_2X_4} - r_{X_1X_4}r_{X_2X_3})]\end{matrix}} \tag{3.65}$$

We see that there are two roots of the equation, whereas we only want one solution. If the assumption that the reliability does not change is correct, the two roots should be the same. If the reliability changes, however, the roots will be different, and we have exactly the same situation we had in the discrete case. We can now understand the presence of these extra roots better, however, because we can see that they correspond to the different times and allow for the possibility that the latent probabilities or reliabilities are changing. There are no roots corresponding to the initial and terminal time points, but for the discrete case there are two roots (a_1 and a_2) and for the continuous case one root corresponding to each time point in between the initial and final waves.

In the present case, if r_{r2} and r_{r3} are different, the equation (3.64) has two different unknowns instead of one. This creates some additional problems, which have not been fully solved, but which will be dealt with again in Chapter 5.

As an example, we have taken the level of aspiration data used in the previous example, and applied the four-wave model to it.

The two solutions for r_r are 0.888 and 0.990. The first root is the more plausible, but it is worth nothing that the two roots correspond rather closely to the separate estimates for r_{r2} and r_{r3} in the previous model, which were about 0.90 and 0.97 respectively. However, the closeness of the new estimates to the earlier ones seems to confirm the use of the simpler model for this set of data.

As for generalizations to m waves, the models have not been worked out, but intuitively it appears that a simple extension of the method of solution, using determinants of higher rank, is all that is required.

CHAPTER 4

Models Combining Fixed Latent Probabilities with Systematic Latent Change

The aggregate mathematical models of econometricians, as they stand, are not directly applicable to panel data. The models based on controlled learning experiments are handling data so different from panel field data that there is no more reason for expecting them to be useful than for expecting any other mathematical system to be useful.

There is, however, a literature proposing models addressed specifically to panel data (Anderson, 1951, 1954) and it is useful to describe the simplest of these models. The discussion will be very brief and elementary, however, because Anderson's own account of the models is easily accessible to the reader who wishes more information.

The basic mathematics of Anderson's simplest model and of most of his models is what is called a Markov chain (Feller, 1957 and many others). The fundamental concept of Markov chains is that of a probability of change (or, complementarily, of not changing). Suppose that there are two positions which a person may occupy at a given time and at some later time. In the case of attitude items, for example, he might be, say, a Republican or a Democrat at a certain time. Suppose he is a Republican, at that time. Then at a given later time, there is a certain probability that he will be a Democrat, and of course a certain probability that he will remain a Republican. If he is a Democrat at the first time, there is a certain probability, which may be the same as or different from that for the Republican, of being a Republican at the second time, and a complementary probability of being a Democrat.

Furthermore, at the first time there is a certain probability of occupying a given position. The person has a certain probability of being a Republican and a complementary probability of being a Democrat. Suppose we want to know the probability that a person will be a Democrat both times. We multiply the probability that he will be a Democrat at the first time by the probability that he will remain a Democrat from time 1 to time 2.

The probabilities of occupying certain positions at time 1 are called the initial distribution. The probabilities of changing are called transition probabilities. If we use the notation

$p_{1,1}$ = the probability the person will be a Democrat at time 1
$p_{2,1}$ = the probability the person will be a Republican at time 1
and
t_{12} = the probability of changing from Democrat to Republican
t_{11} = the probability of remaining Democratic
t_{21} = the probability of changing from Republican to Democratic
t_{22} = the probability of remaining Republican
we can define a vector

$$p_1 = \begin{pmatrix} p_{1,1} \\ p_{2,1} \end{pmatrix}$$

as the initial distribution.

We can also define a matrix

$$T = \begin{pmatrix} t_{11} & t_{12} \\ t_{21} & t_{22} \end{pmatrix}$$

as the matrix of transition probabilities, in which each row adds up to one.

Now if we want to predict the distribution at time 2, it is merely necessary to take the probability for each position at time 1 in turn and multiply it by the probability of going to a given position, then add these products. For example, the probability of being a Democrat at time 2 is

$$p_{1,2} = p_{1,1}t_{11} + p_{2,1}t_{21}$$

or in matrix notation

$$p'_2 = p'_1 T$$

There are two more concepts required to complete the definition of the simplest model — "order of the process" and "stationary transition probabilities". The concept of order of a process is one which we used in Chapter 2 and 3 even when we were not talking about Markov chains. The order of a process means the number of times prior to a given time on which the position held at that given time depends. Thus a first-order process would be one in which the position held at time 3 depends only on the position held at time 2 and not on the position held at time 1. It is important to realize that this does not mean that the position at time 3 is unrelated to the position at time 1. Putting it in terms more familiar to the survey analyst, it means that if the position at time 2 is controlled the position at time 3 is unrelated to the position at time 1. Similarly, a second-order process would be one in which the position at time 4 is not unrelated to the position at time 2 when the position at time 3 is controlled, but is unrelated to the position at time 1 when the positions at time 2 and time 3 are controlled.

Anderson's simplest model is a straightforward first order process. He then goes on to show that second and higher order processes can always be formulated as first order processes, by defining a given position at time 1 in terms of the joint positions held at that and the previous or several previous times. For example, suppose we want to convert a second order process into a first order process, in the case cited. We say that there are four positions at time 1: having been a Democrat at time 0 and still being a Democrat, having been a Republican and now being a Democrat, having been a Democrat and now being a Republican, and having been a Republican and now being a Republican. Each one of these joint positions is called a state. Then the states can be handled as the positions were in the first order case.

The concept of stationary transition probabilities means that the matrix of transition probabilities does not change through time. If a person's vote intention has stationary transition probabilities, then if he were a Republican in May his probability of becoming a Democrat in June would be the same as the probability that if he were a Republican in September he would become a Democrat in October. The models which Anderson proposed for panel data all have stationary transition probabilities. First order processes with stationary transition probabilities are called Markov chains, after their inventor.

Now how, in practice, are the probabilities determined? Suppose that instead of one individual we have a sample of many individuals, who are homogeneous in the sense that they all follow the same process. We can then use the proportions of people who occupy a given position or who change from a given position to another as estimates of the probabilities, subject of course to sampling variation in application.

Given the definitions and assumptions of the process, a great many interesting theorems predicting the future of the system can be derived, as Anderson shows. In particular, it can be easily shown that in a Markov chain the probabilities of going from any particular position to any other particular position from t to time $t + m$ are given by T^m, the mth power of the matrix of transition probabilities. If the distribution at any one time is known, and if the matrix of transition probabilities is known, then the distribution at any time in the future can be predicted. Suppose that m is very large, and we compute the matrix T^m. Consider then the matrix for the time interval from t to $t + m + 1$, which is equal to T^{m+1}. Since m is very large T^m and T^{m+1} are going to be almost identical, and as m approaches infinity they will become equal. That means that the distribution at time $t + m$ and the distribution at time $t + m + 1$ will be alike, and the distribution at all succeeding times will be the same. When the distribution gets to this point it is called a stationary distribution or equilibrium state. In terms of panel data, it means that the marginals as well as the turnover remain constant through time.

This extremely brief nontechnical description of Anderson's simplest model gives the essentials of the model. In commenting on this model, we should make clear that there is no interest in attacking or defending it. Whether the model fits data is a matter for empirical determination. There are, nonetheless, several reasons why Markov chain models, taken by themselves, are not promising for the analysis of panel data, especially attitude items (Wiggins, 1951, 1955a; the first chapter of Coleman, 1964, has a good resume of this situation).

In the first place, as will be further discussed in Chapter 5, not a single set of data studied here showed a first order process, and it is likely that the lowest order to be found in panel data of attitudes is third or fourth order, perhaps even higher. Now although it is possible in principle to deal with third or fourth order processes, these require even for dichotomous items sixteen or thirty-two separate frequencies to serve as bases for the estimates of transition probabilities. To provide decent minimum numbers for estimation would require very large samples. Furthermore, three or four time points would be required for estimation, and few panel studies can afford four or five waves with very large samples. From a practical point of view, then, we cannot expect to find these models very useful.

In the second place, if the phenomena which the models proposed in this monograph attempt to describe are found in a given set of data — and it is argued that they are for most sets of data with limited controls for example, field data — then the Markov chain cannot hold *for the manifest data.* It is still possible that the Markov chain model may hold for the latent positions, as these have been defined. In this chapter, there is a brief discussion of how the latent probability models (the models we propose) may be combined with a Markov chain, with the solution for one simple case.

The third and perhaps most cogent reason for questioning the utility of the Markov chain model for attitude data is that it treats a single attitude as a closed system, making it completely dependent on its own past. Anderson also presents a model for two or more items, which is subject to essentially the same criticisms. This is certainly a false assumption for peripheral attitudes and behavior of the kind usually studied, since we know that they are subject to the influence of other attitudes and of exogenous influences of all sorts. On the other hand, the more autonomous central attitudes and traits hardly ever change, so that they might constitute a special but trivial case of a Markov chain where the transition matrix would be an identity matrix.

If models are to be constructed for single attitudes changing through time, it appears to the writer unpromising to construct them so that they view the attitude as a closed system, that is, to specify completely what happens or what will happen. The best models presented here are incomplete in a sense, because most of them admit a number of parameters which are unexplained or exogenous, but it is perhaps for this very reason that they are able to fit certain bodies of data rather closely.

To repeat, if latent probability models of the type we have been considering will fit a wide range of panel data — and we believe that they will — then Markov chain models based on a stationary process will not fit the manifest data. If, in other words, there is a substantial random component in a measured attitude or behavior at each time point, whether error of measurement, randomness in the behavior or attitude, or both, the Markov chain model by itself is not adequate.

This conclusion is supported by examination of all the empirical data available to the writer, which included hundreds of cases where a discrete variable was measured at three time points, in scores of different panel studies, and two cases where a discrete variable was measured at six time points, from the Sandusky study. For the three-wave data, in no case was the variable at time 3 independent of itself at time 1 when time 2 was controlled, in the manifest data. Thus no manifest first-order process was found.

Further, there is no reason to think that a second order process would fit the data. Two examples with more than three time points were examined; these were "vote intention" and "winner expectation" in the 1940 political study which we have referred to before. It was the writer's task to try to locate examples for Anderson's original paper (1951), and the Sandusky data were investigated for the purpose. The "winner expectation" data, which were used by Anderson, do not conform to a first order process, as Anderson himself points out. For that matter, they do not conform to a second order process either.

If the relationship between vote intention at a given time and vote intention at the third previous wave is examined, holding constant the vote intention at the two intermediate waves, we find that there is still a dependence. The following table shows the relationships which are found, where the variable is trichotomized at the intermediate times into the classes "Democratic", "Republican", and "Don't know", and dichotomized at the terminal and initial times into "Republican" and "Not Republican", and where the phi-coefficient is used as the measure of association (see Table 10.)

Altogether, there are eleven positive relationships, four negative, one equal to zero, and eleven indeterminate (indeterminate means that one row or column of the four-fold table is zero) out of twenty-seven relationships. Those surrounded with parentheses are based on four-fold tables with fewer than ten cases in the table. If these are eliminated (fourteen in number), there are nine positive, two negative, one zero, and one indeterminate. While the number of comparisons which can be made is small, it is highly likely that the position at a given time is dependent on the position at the

TABLE 10

Signs of partial associations of vote intention with itself at the third previous wave, controlling vote intention at the first and second previous wave
(Data from 1940 voting study in Sandusky, Ohio. $N = 343$)
Sign of association between time i and time $i-3$

Position at time $i-2$	*Position at time $i-1$*	*$i-3$ = May i = August*	*$i-3$ = June i = September*	*$i-3$ = July i = October*
Republican	Republican	Positive	Negative	Positive
Republican	Democratic	(Positive)	(Positive)	(Negative)
Republican	Don't know	Zero	Positive	(Indeterminate)
Democratic	Republican	(Indeterminate)	(Indeterminate)	(Indeterminate)
Democratic	Democratic	Negative	Positive	Indeterminate
Democratic	Don't know	(Indeterminate)	Positive	(Indeterminate)
Don't know	Republican	Positive	(Indeterminate)	(Indeterminate)
Don't know	Democratic	(Indeterminate)	(Indeterminate)	(Negative)
Don't know	Don't know	Positive	Positive	Positive

third previous time even when the positions at the two intermediate times are controlled.

Further, there is no reason to think the dependency stops at the third previous time. We simply cannot trace it further back because we run out of cases. We may summarize the results of analysis of the available bodies of empirical data. then, by saying that in the manifest data, for discrete variables, we have not been able to find a case in which the terminal position is independent of the initial position, when the intermediate positions are controlled. Kuehn (1958) got the same dependency in studies of purchases of frozen orange juice. I do not know of any counterexample in the literature.

Of course, it is possible to convert almost any process into a stationary process by telescoping enough past time points into one. This is a pleasant pursuit for mathematicians, but a jejune endeavor for most attitude and market research workers, since in practice the number of time points and the number of cases is almost never sufficient. If we have a third order process involving a trichotomous variable, we need twenty-seven classes in order to predict the future, which in most studies will reduce the number of available cases per class to around fifteen or twenty, as bases for estimates of probabilities.

However, the fact that the manifest data rarely seem to fit models of stationary Markov processes of low order does not mean that such models may not hold for the latent positions or latent classes. In this chapter we want to look very briefly at the possibility of combining the latent probability model with the first order Markov chain model, in terms of the appearance of the manifest data and the solution for the parameters.

Let us recall that a stationary process means that there is a certain probability that a person holding any given position at some time t will hold any given position at time $t + u$, and that this probability depends on u but not on t. In other words, the interval between the two times is all that matters. Once the size of the interval is fixed, the probability of going from one given position to another over this interval is the same regardless where this interval falls in the series, hence the term "stationary process". Probabilities of this type are called transition probabilities.

If there is a stationary process such that the position held at a given interview depends only on the position held the previous interview, the process is said to be of the first order. If the position held depends on both the previous position and the position held two times previously, the process is of the second order, and so on.

We want to investigate briefly the relationship between the manifest data and latent process when the item involves a random component. How will manifest data look if there is a latent stationary process?

Let us consider a first order stationary process. Such a process implies two things about any sequence of three successive interviews from a panel where the interviews are spaced at equal time intervals apart. It implies, first, that the transition probabilities from time 1 to time 2 are the same as the corresponding transition probabilities from time 2 to time 3. It implies, secondly, that the position held at time 3 is independent of the position held at time 1, depending only on the position at time 2.

Since the equality of the successive transition probabilities is not a sufficient condition for a first order stationary process let us introduce a terminological distinction and call such equality "stationary probabilities". Then in order to have a first order stationary process, we need first order stationary probabilities and the independence of the position at time 3 from the position at time 1.

Now suppose we have true first order stationary probabilities, with nothing said about the independence condition. Let v_{ijk} be the proportion of the total sample in class C_i at time 1, in class C_j at time 2, and in class C_k at time 3, as before. Then $v_{11.}/v_{1..}$ will be the probability of going from C_1 at time 1 to C_1 at time 2; and $v_{.11}/v_{.1.}$ will be the probability of going from C_1 at time 2 to C_1 at time 3. By assumption, these probabilities are equal, or

$$\frac{v_{11.}}{v_{1..}} = \frac{v_{.11}}{v_{.1.}}$$

Similarly,

$$\frac{v_{12.}}{v_{1..}} = \frac{v_{.12}}{v_{.1.}}$$

Combining these two relations,

$$\frac{v_{11.}}{v_{12.}} = \frac{v_{.11}}{v_{.12}}$$

or, in determinantal form,

$$\begin{vmatrix} v_{11.} & v_{.11} \\ v_{12.} & v_{.12} \end{vmatrix} = 0 \tag{4.1}$$

By the same reasoning, we get also

$$\begin{vmatrix} v_{21.} & v_{.21} \\ v_{22.} & v_{.22} \end{vmatrix} = 0 \tag{4.2}$$

Now consider the following array of manifest data:

$$\begin{matrix} 1 & 1 & \quad & p_{1..} & p_{.1.} \\ p_{.1.} & p_{..1} & \quad & p_{11.} & p_{.11} \end{matrix}$$

Let us set up the determinant of the first square of this array in terms of the latent structure, as follows:

$$\begin{vmatrix} 1 & 1 \\ p_{.1.} & p_{..1} \end{vmatrix} = \left| \begin{pmatrix} 1 & 1 & 1 & 1 \\ a_1 & a_2 & a_1 & a_2 \end{pmatrix} \begin{pmatrix} v_{11.} & v_{.11} \\ v_{12.} & v_{.12} \\ v_{21.} & v_{.21} \\ v_{22.} & v_{.22} \end{pmatrix} \right| = |AV| \tag{4.3}$$

This equation is based on our present restriction to two latent classes at any one time and two manifest response categories. It follows clearly from our basic assumptions.

The expansion of the right hand side of (4.3), which we shall call $|AV|$, is the sum of six products of minors of order 2 from columns of A and minors from the corresponding rows of V. By (4.1) and (4.2), the first and last terms of the expansion will vanish.

We next put the determinant of the second square of manifest data in terms of the latent structure:

$$\begin{vmatrix} p_{1..} & p_{.1.} \\ p_{11.} & p_{.11} \end{vmatrix} = \left| \begin{pmatrix} 1 & 1 & 1 & 1 \\ a_1 & a_2 & a_1 & a_2 \end{pmatrix} \begin{pmatrix} a_1 v_{11.} & a_1 v_{11} \\ a_1 v_{12.} & a_1 v_{12} \\ a_2 v_{21.} & a_2 v_{21} \\ a_2 v_{22.} & a_2 v_{22} \end{pmatrix} \right| \tag{4.4}$$

Comparing (4.4) with (4.2), observe that on the right hand side the only difference is that in (4.4) the first two rows of V have been multiplied by a_1 and the last two rows by a_2. In expanding the right hand side of (4.4), the first and last product in the sum will vanish because the first minor of the second rectangular matrix in (4.4) is the determinant of (4.1) multiplied by a_1^2 and the last minor of this matrix is the determinant of (4.2) multiplied by a_2^2. The remaining terms of the expansion will all involve a minor of V with one row multiplied by a_1 and the other multiplied by a_2. Hence $a_1 a_2$ can be factored out, and what remains is the expansion of $|AV|$, or

$$\begin{vmatrix} p_{1..} & p_{.1.} \\ p_{11.} & p_{.11} \end{vmatrix} = a_1 a_2 |AV| \tag{4.5}$$

Dividing (4.5) by (4.3),

$$\frac{\begin{vmatrix} p_{1..} & p_{.1.} \\ p_{11.} & p_{.11} \end{vmatrix}}{\begin{vmatrix} 1 & 1 \\ p_{.1.} & p_{..1} \end{vmatrix}} = a_1 a_2 \tag{4.6}$$

Now form two matrices by substituting one column of the second square of the initial array of manifest data for the corresponding column of the first square of the array. The determinants of these two matrices may be put in terms of the latent structure as follows:

$$\begin{vmatrix} 1 & p_{.1.} \\ p_{.1.} & p_{.11} \end{vmatrix} = \left| \begin{pmatrix} 1 & 1 & 1 & 1 \\ a_1 & a_2 & a_1 & a_2 \end{pmatrix} \begin{pmatrix} v_{11.} & a_1 v_{.11} \\ v_{12.} & a_1 v_{.12} \\ v_{21.} & a_2 v_{21.} \\ v_{22.} & a_2 v_{22.} \end{pmatrix} \right| \tag{4.7}$$

and

$$\begin{vmatrix} p_{1..} & 1 \\ p_{11.} & p_{..1} \end{vmatrix} = \left| \begin{pmatrix} 1 & 1 & 1 & 1 \\ a_1 & a_2 & a_1 & a_2 \end{pmatrix} \begin{pmatrix} a_1 v_{11.} & v_{.11} \\ a_1 v_{12.} & v_{.12} \\ a_2 v_{21.} & v_{.21} \\ a_2 v_{22.} & v_{.22} \end{pmatrix} \right| \tag{4.8}$$

In expanding the right hand side of each of these equations, note that the first and last product will vanish, because in each case the first and last product will involve a vanishing minor of V with one column multiplied by a_1 or a_2, which can be factored out. If we add the remaining terms of (4.7) and (4.8), collecting for minors of A, we get terms of the following type:

$$\begin{vmatrix} 1 & 1 \\ a_i & a_j \end{vmatrix} \left\{ \begin{vmatrix} v_{1i.} & a_1 v_{.1i} \\ v_{2j.} & a_2 v_{.2j} \end{vmatrix} + \begin{vmatrix} a_1 v_{1i.} & v_{.1i} \\ a_2 v_{2j.} & v_{.2j} \end{vmatrix} \right\}$$

By the same reasoning employed previously, the bracketed expression is equal to

$$(a_1 + a_2) \begin{vmatrix} v_{1i.} & v_{.1i} \\ v_{2j.} & v_{.2j} \end{vmatrix}$$

Hence $(a_1 + a_2)$ can be factored out of the expansion, and what remains is the expansion of $|AV|$, or

$$\begin{vmatrix} 1 & p_{.1.} \\ p_{.1.} & p_{.11} \end{vmatrix} + \begin{vmatrix} p_{1..} & 1 \\ p_{11.} & p_{..1} \end{vmatrix} = (a_1 + a_2)|AV| \tag{4.9}$$

Dividing (4.9) by (4.3)

$$\frac{\begin{vmatrix} 1 & p_{.1.} \\ p_{.1.} & p_{.11} \end{vmatrix} + \begin{vmatrix} p_{1..} & 1 \\ p_{11.} & p_{..1} \end{vmatrix}}{\begin{vmatrix} 1 & 1 \\ p_{.1.} & p_{..1} \end{vmatrix}} = a_1 + a_2 \tag{4.10}$$

Finally, therefore, a_1 and a_2 are the roots of the equation

$$x^2 - \frac{\begin{vmatrix} 1 & p_{.1.} \\ p_{.1.} & p_{.11} \end{vmatrix} + \begin{vmatrix} p_{1..} & 1 \\ p_{11.} & p_{..1} \end{vmatrix}}{\begin{vmatrix} 1 & 1 \\ p_{.1.} & p_{..1} \end{vmatrix}} x + \frac{\begin{vmatrix} p_{1..} & p_{.1.} \\ p_{11.} & p_{.11} \end{vmatrix}}{\begin{vmatrix} 1 & 1 \\ p_{.1.} & p_{..1} \end{vmatrix}} = 0 \qquad (4.11)$$

Once a_1 and a_2 are obtained, it is easy to find the proportions in the various true classes. Note that (4.11) does not involve $p_{1.1}$ or p_{111}.

What we have obtained is a solution for the probabilities of positive response on the assumption of latent first order stationary probabilities. A latent first order stationary process would require this assumption, and also the assumption that latent position at time 3 is independent of the latent position at time 1. In Chapter 3 we obtained a solution for a_1 and a_2 on the basis of the independence assumption alone, where we had

$$\frac{\begin{vmatrix} 1 & p_{.11} \\ p_{1..} & p_{111} \end{vmatrix} + \begin{vmatrix} p_{.1.} & p_{..1} \\ p_{11.} & p_{1.1} \end{vmatrix}}{\begin{vmatrix} 1 & p_{..1} \\ p_{1..} & p_{1.1} \end{vmatrix}} = a_1 + a_2 \qquad (4.12)$$

and

$$\frac{\begin{vmatrix} p_{.1.} & p_{.11} \\ p_{11.} & p_{111} \end{vmatrix}}{\begin{vmatrix} 1 & p_{..1} \\ p_{1..} & p_{1.1} \end{vmatrix}} = a_1 a_2 \qquad (4.13)$$

If there is a latent first order stationary process, the solution for a_1 and a_2 on the basis of the assumption of latent first order stationary probabilities and the solution on the basis of the assumption of the independence of the latent position at time 3 from the latent position at time 1 must be the same, because both of these assumptions must hold if there is a latent first order stationary process. Hence (4.10) must be equal to (4.12), and (4.6) must be equal to (4.13). This gives us the following two conditions which the manifest data must fulfill if there is a latent first order stationary process:

$$\frac{\begin{vmatrix} 1 & p_{.1.} \\ p_{.1.} & p_{.11} \end{vmatrix} + \begin{vmatrix} p_{1..} & 1 \\ p_{11.} & p_{..1} \end{vmatrix}}{\begin{vmatrix} 1 & 1 \\ p_{.1.} & p_{..1} \end{vmatrix}} = \frac{\begin{vmatrix} 1 & p_{.11} \\ p_{1..} & p_{111} \end{vmatrix} + \begin{vmatrix} p_{.1.} & p_{..1} \\ p_{11.} & p_{1.1} \end{vmatrix}}{\begin{vmatrix} 1 & p_{..1} \\ p_{1..} & p_{1.1} \end{vmatrix}} \tag{4.14}$$

$$\frac{\begin{vmatrix} p_{1..} & p_{.1.} \\ p_{11.} & p_{.11} \end{vmatrix}}{\begin{vmatrix} 1 & 1 \\ p_{.1.} & p_{..1} \end{vmatrix}} = \frac{\begin{vmatrix} p_{.1.} & p_{.11} \\ p_{11.} & p_{111} \end{vmatrix}}{\begin{vmatrix} 1 & p_{..1} \\ p_{1..} & p_{1.1} \end{vmatrix}} \tag{4.15}$$

If the manifest data fulfill these conditions for every set of three successive interviews, the assumption of a latent first order stationary process is warranted, and the latent structure may be obtained by (4.11) or by the equations of Chapter 3.

Let us investigate further the relation between latent first order stationary process and manifest data. Let us define P_i as a diagonal matrix of the manifest marginal relative frequencies at time i. Define P_{ij} as the matrix of manifest joint relative frequencies at time i and time j. Let T_{ij} be the matrix of manifest transition probabilities from time i to time j. Then

$$P_i^{-1} P_{ij} = T_{ij}$$

Similarly define V_i as a diagonal matrix of the latent marginal relative frequencies at time i, let V_{ij} be the matrix of latent joint relative frequencies at time i and time j, and let H_{ij} be the matrix of latent transition probabilities from time i to time j. Then

$$H_{ij} = V_i^{-1} V_{ij}$$

Now let $A = (a_{ij})$ be the matrix of probabilities that a person in latent class C_i will give a response falling in the manifest category j. Then from our basic assumptions we have the equations

$$P_i = A' V_i$$

and

$$P_{ij} = A' V_{ij} A$$

Premultiplying the second equation by the identity $P_i P_i^{-1}$ on the left hand side and inserting between A' and V_{ij} the identity

$V_i V_i^{-1}$ on the right hand side,

$$P_i P_i^{-1} P_{ij} = A' V_i V_i^{-1} V_{ij} A$$

By the equations above, this is

$$P_i T_{ij} = A' V_i H_{ij} A \qquad (4.16)$$

Premultiplying both sides by P_i^{-1},

$$T_{ij} = P_i^{-1} A' V_i H_{ij} A \qquad (4.17)$$

If we consider first order transition probabilities for three interviews, we have by (4.17) two equations:

$$T_{12} = P_1^{-1} A' V_1 H_{12} A$$

and

$$T_{23} = P_2^{-1} A' V_2 H_{23} A$$

Now if we have true first order stationary probabilities,

$$H_{12} = H_{23}$$

Therefore, subtracting the second of the two equations from the first,

$$\begin{aligned} T_{12} - T_{23} &= P_1^{-1} A' V_1 H_{12} A - P_2^{-1} A' V_2 H_{23} A \\ &= (P_1^{-1} A' V_1 - P_2^{-1} A' V_2) H_{12} A \qquad (4.18) \end{aligned}$$

Now if the manifest first order probabilities are stationary,

$$T_{12} = T_{23}$$

or

$$T_{12} - T_{23} = (0)$$

By (4.18) this is the case if and only if

$$P_1^{-1} A' V_1 - P_2^{-1} A' V_2 = (0)$$

or

$$P_1^{-1} A' V_1 = P_2^{-1} A' V_2 \qquad (4.19)$$

Hence if the latent first order transition probabilities are stationary, (4.19) is the necessary and sufficient condition for the manifest first order transition probabilities to be stationary.

(4.19) evidently holds in the case of perfect reliability, for then $A' = I$, $P_1 = V_1$, and $P_2 = V_2$, so that each side of (4.19) becomes I. (4.19) also holds if

$$P_1 = P_2$$

and

$$V_1 = V_2$$

i.e., if the latent and manifest marginals are both unchanging from time 1 to 2. Now if $V_1 = V_2$, we have

$$P_1 = A'V_1 = A'V_2 = P_2$$

so that (4.19) holds if the latent marginals are the same at times 1 and 2. Therefore a sufficient condition for manifest first order stationary probabilities is

$$V_1 = V_2 \tag{4.20}$$

and

$$H_{12} = H_{23}$$

In other words, when we have a latent stationary process in the stationary state (which means that V_1 will be equal to V_2) (Anderson, 1951), we will have a manifest stationary process. In general, however, in the case of latent probabilities less than one, a latent stationary process not in the stationary state will not give rise to a manifest stationary process.

This whole relationship could profitably be explored further. Lazarsfeld and Henry (1968) have produced some additional results on this topic, but much remains to be done.

CHAPTER 5

Latent Probability Models with Changing Latent Probabilities and no Latent Change

To this point we have considered only models in which the latent probabilities remain constant throughout time. We now turn to consideration of situations in which the latent probabilities change, either under restrictions or completely freely. It turns out that these models, in one form or another, fit a surprisingly large proportion of the data examined, that is, panel data where three or more waves are available, since there is a huge mass of data where only two waves are employed. Despite this, the range of applicability of the models will be seen to be quite wide.

For this reason we shall lay more emphasis on examples in this and the following chapter than in the others. We shall also apply the models in such a way as to leave some degrees of freedom in the system in every case, so that although statistical tests will not be applied, we can get at least an intuitive sense of the "goodness of fit" of these models. In other words, any fits that we get will not be trivial solutions in which the number of unknowns equals the number of items of information.

The point in this procedure is that from these applications we shall be able to make a few empirical generalizations about what might loosely be called social or psychological process. The literature is surprisingly weak in providing determinate descriptive generalizations about these processes of attitude and behavior, most of the propositions about such processes being directed at the conditions under which change may be expected to occur rather than to the precise descriptions of the phenomenon itself. We are concerned here with what is happening and with providing a seemingly reasonable explanation of how it is happening.

We consider here models in which the data are completely accounted for by the latent probabilities, without invoking changes from one latent class to another. In the present models, unlike those of previous chapters, the latent probabilities can change. We might impose some restriction on this change, such as linearity, but this seems somewhat arbitrary, particularly since the movement of the latent probabilities is a matter of interest in its own right. The restriction we shall use will be that in a number of models, in order to simplify solutions, and where the manifest data seem clearly to warrant the assumption, we shall assume that over some particular time interval the latent probabilities do not change. The reason for this assumption will become clearer as we proceed.

We shall begin with an example of a solution for two waves, a solution which is trivial in the sense that there are no degrees of freedom. Thereafter, we shall deal with three time points, where the models do not have to fit.

Two Waves, Two Latent Probabilities

We shall change the notation a little bit to simplify the equations for this chapter. Let us call the proportion giving the positive manifest response at time i "p_i", and the proportion giving the positive response at both times i and j "p_{ij}". We shall continue to use the notation v_{ijk} to mean the proportion in the latent class i at time 1, j at time 2, and k at time 3. Since all the models of this chapter involve only two latent classes at any one time, and only two possible responses at any one time, we can unambiguously use the notation a_i to refer to the probability that a person in latent class 1 at time i will give the positive response at time i. It happens that in all the models which will be shown we assume that the two latent classes correspond to the two manifest classes and that the probability is the same for each latent class that the person will be placed by the manifest data in his corresponding position. This means that if the probability at time 1 is a_1 for a person in the latent class 1 to give the positive response (or be placed in the positive category) then the probability is also a_1 for the person in latent class 2 to give the negative response. Thus the probability is $(1 - a_1)$ that a person in latent class 2 at time 1 will give the positive response at time 1.

Consider the cases where there are two time points, no changes from one latent class to another, but the probability a_1 may differ from a_2. The parameters of the model are as follows:

Time	*Latent probability*	*Proportion in latent class 1*	*Proportion in latent class 2*
1	a_1	v_{11}	v_{22}
2	a_2	v_{11}	v_{22}

The manifest data include p_1, the proportion positive at time 1, p_2, the proportion positive at time 2, and p_{12}, the proportion positive at both times.

We have, by the basic assumptions of the model,

$$p_1 = a_1 v_{11} + (1-a_1)v_{22} \tag{5.1}$$

$$p_2 = a_2 v_{11} + (1-a_2)v_{22} \tag{5.2}$$

$$p_{12} = a_1 a_2 v_{11} + (1-a_1)(1-a_2)v_{22} \tag{5.3}$$

Multiplying (5.1) by (5.2),

$$p_1 p_2 = a_1 a_2 v_{11}^2 + a_1(1-a_2)v_{11}v_{22} + a_2(1-a_1)v_{11}v_{22} + (1-a_1)(1-a_2)v_{22}^2 \tag{5.4}$$

Hence

$$\begin{aligned}[12] &= P_{12} - p_1 p_2 \\ &= a_1 a_2(v_{11}^2 - v_{11}) + (1-a_1)(1-a_2)(v_{22}^2 - v_{22}) \\ &\quad - a_1(1-a_2)v_{11}v_{22} - a_2(1-a_1)v_{11}v_{22}\end{aligned} \tag{5.5}$$

Using the fact that $v_{22} = 1 - v_{11}$, collecting terms, this gives

$$[12] = (2a_1 - 1)(2a_2 - 1)v_{11}(1 - v_{11}) \tag{5.6}$$

From (5.1),

$$p_1 = a_1 v_{11} + (1-a_1)(1-v_{11})$$

and

$$1 - v_{11} = 1 - \frac{p_1 + a_1 - 1}{2a_1 - 1}$$

$$1-v_{11}=\frac{a_1-p_1}{2a_1-1} \tag{5.7}$$

From (5.2),

$$v_{11}=\frac{a_2+p_2-1}{2a_2-1} \tag{5.8}$$

Substituting (5.7) and (5.8) in (5.6),

$$[12]=(a_2+p_2-1)(a_1-p_1) \tag{5.9}$$

This equation involves the unknowns a_1 and a_2. From (5.7) and (5.8) we can derive the equation

$$a_1=\frac{p_1(2a_2-1)+p_2-a_2}{2p_2-1} \tag{5.10}$$

This equation is satisfactory unless $p_2 = 0.50$, in which case the equation is indeterminate. Equations (5.9) and (5.10) can otherwise be solved for a_1 and a_2, giving the solutions

$$a_1=0.5\pm\sqrt{0.25-p_1(1-p_1)+[12]\frac{(2p_1-1)}{(2p_2-1)}} \tag{5.11}$$

$$a_2=0.5\pm\sqrt{0.25-p_2(1-p_2)+[12]\frac{(2p_2-1)}{(2p_1-1)}} \tag{5.12}$$

We can then use (5.8) to solve for v_{11}.

Example of Two-Wave Model

We shall take as an example data from the study of the 1948 election campaign in Elmira, New York (see Berelson *et al.*, 1954). Using only data from the interviews in August and October, and dichotomizing the responses into "Republican" and "Not Republican", the data are given in Table 11.

From this we get

$p_1 = 0.664$

$p_2 = 0.667$

$p_{12} = 0.627$

$[12] = 0.185$

TABLE 11

Vote intention, August and October, Elmira Panel

		October		
		Republican	Not Republican	Total
August	Republican	353	20	373
	Not Republican	22	167	189
	Total	375	187	562

Working out the solutions gives the following structure:

Time	*Latent probability*	*Proportion in latent class 1*	*Proportion in latent class 2*
1	0.956	0.686	0.314
2	0.966	0.686	0.314

It is useful to carry out all computations to four significant digits, and results will generally be reported to three places. We see that 0.686 of the sample are "True Republicans," and that the latent probability that a respondent in a given latent class will be found in the corresponding manifest class is quite high at both times, going up by 0.01 from August to October.

Three Waves, One Change in Latent Probabilities

Inspection of the data often shows that the amount of turnover or change between the first and third waves of a three wave panel is about the same as that between the first and second waves, but the amount of turnover is less between the second and third waves. In such situations it is not unreasonable to assume that the latent probability is the same at times 2 and 3, but different at time 1. In such cases, we have the following latent probability structure:

Time	*Latent probability*	*Proportion in latent class 1*	*Proportion in latent class 2*
1	a_1	v_{111}	v_{222}
2	a_2	v_{111}	v_{222}
3	$a_3 = a_2$	v_{111}	v_{222}

The model as it applies to waves 2 and 3 is the same as the first and simplest latent probability model proposed in Chapter 2, with two latent classes and no latent change. From that model we get the solution for $a_2 = a_3$ and for v_{111} and v_{222}, as follows:

$$a_2 = a_3 = 0.5 + \sqrt{0.25 + p_{23} - p_2}$$

$$v_{111} = \frac{p_2 + a_2 - 1}{2a_2 - 1}$$

The assumption that neither the latent probabilities nor latent proportions change from time 2 to 3 of course implies that the manifest marginals do not change, or that $p_2 = p_3$.

Once a_2 and v_{111} and v_{222} are known, it is easy to find a_1, the remaining unknown.

$$p_{12} = a_1 a_2 v_{111} + (1-a_1)(1-a_2)v_{222}$$

$$= a_1(a_2 v_{111} + a_2 v_{222} - v_{222}) + v_{222} - a_2 v_{222}$$

$$a_1 = \frac{p_{12} - (1-a_2)v_{222}}{a_2 v_{111} - (1-a_2)v_{222}} \qquad (5.13)$$

In this solution, as in all the solutions of this chapter, we do not use the manifest third order frequencies p_{123}. We therefore always have at least one degree of freedom left in fitting the data, since there is one degree of freedom associated with the third order frequency; that is, the numerical value of the third order frequency is not determined (although certain limits are determined for it) by the first and second order frequencies.

Example of the Model with Three Waves and One Change in Latent Probabilities

This example is drawn from behavior rather than attitude. As part of the data in a number of studies of medical students and the medical profession, information was collected about the grades made by all the students in a given medical school during their first three years in the school. The data were kindly furnished to the writer by Dr. Patricia Kendall Lazarsfeld from one of the medical studies under the general direction of Dr. Robert K. Merton. These grades were summary grades for an entire year of study in each case, and were awarded by a faculty committee on the basis of all course and examination grades made during the year. Hence it is likely that the reliability of grading was very high. The faculty divided the students into four approximately equal grade groups at the end of each year, and for our purposes these groups have been collapsed into two at the point nearest the median. As a result of the method of assignment of the grades, the marginal proportion "high" does not change at all. Looking at the three waves pairwise, the data are as follows:

TABLE 12

Grade patterns of medical students over first three years, by pairs of years

		1st year		
		High	Low	
2nd Year	High	56	24	80
	Low	24	52	76
		80	76	156

		1st year		
		High	Low	
3rd Year	High	55	25	80
	Low	25	51	76
		80	76	156

		2nd year		
		High	Low	
3rd Year	High	61	19	80
	Low	19	57	76
		80	76	156

The first thing to observe is the feature of data of this type which may seem surprising, namely, that the amount of turnover from the first to the third year is no greater than that from the first to the second year. From the second to the third year, however, the turnover is a good deal smaller.

We find that the manifest proportions are as follows:

$p_1 = p_2 = p_3 = 0.512$

$p_{12} = 0.359$

$p_{13} = 0.352$

$p_{23} = 0.391$

Using p_2 and p_{23}, we find that

$a_2 = a_3 = 0.86$

$v_{111} = 0.51$

$v_{222} = 0.49$

Putting these values into (5.13) gives

$a_1 = 0.77$

The structure, then, is as follows:

Time	*Latent probability*	*Proportion in latent class 1*	*Proportion in latent class 2*
1	0.77	0.51	0.49
2	0.86	0.51	0.49
3	0.86	0.51	0.49

Since we have at least one degree of freedom, we can use this structure to generate the complete table of third order frequencies, and compare the generated data with the observed data. We do this by means of formulas for the third order frequencies derived from the basic assumptions of the model, as follows:

$$p_{123} = a_1 a_2 a_3 v_{111} + (1-a_1)(1-a_2)(1-a_3)v_{222} \tag{5.14}$$

By equations of this type, we can generate the proportion positive at all times, as in (5.14), the proportion positive at times 1 and 2 and negative at time 3, and so on through the eight permutations. Multiplying these proportions by the total number of cases, 156, gives estimates, based on the model, of the eight third order frequencies, which are shown in the table below next to the observed frequencies.

This is an almost perfect fit, and it is the more striking because it is based on a model with only three independent unknowns. It is such a good fit that it makes us consider seriously the implications of the model for the situation. The model implies that there are really two classes of students, the better and the worse groups, each constituting about half of the total number. The actual grades these students receive do not accurately define their excellence, probably because there is a random factor in their performance. What is this random factor? We can think of many more or less fortuitous circumstances which might influence a student's performance, things such as illness, falling in love, family discord, deaths, difficulties in getting oriented to the regimen of medical school — factors which might affect some students one year and others the next. The most important group of random factors probably relates to the floundering around some students do in first getting oriented to a new situation. As they get oriented and the situation becomes psychologically stabilized (or the roles become learned), the random element decreases, and the probability that their grades will classify them accurately increases. Even then, the probability is not perfect. It would be interesting to see whether the probability increases in the fourth and final year.

TABLE 13

Comparison of observed and estimated data on grades of medical students

Grade position at			*Estimated frequency*	*Observed frequency*	*Deviation*
1st Year	*2nd Year*	*3rd Year*			
High	High	High	45.9	45	−0.9
High	High	Low	9.4	11	1.6
High	Low	High	9.4	10	0.6
High	Low	Low	14.1	14	−0.1
Low	High	High	14.8	16	1.2
Low	High	Low	9.3	8	−1.3
Low	Low	High	9.3	9	−0.3
Low	Low	Low	43.8	43	−0.8
		Total	156.0	156	0.0

This model describes the data very well, but we shall next develop a slightly different model which describes them even better.

Three Waves, 0.50 Marginals, Latent Probabilities Unrestricted

This model differs from the previous one in that we do not assume that the latent probabilities are the same at times 2 and 3. We continue to assume that no one changes his latent class, as in all the models of this chapter. We want to consider here the case in which the manifest marginals, or proportions giving the positive response, are all the same (or virtually the same) and equal to (or almost equal to) 0.50. As a matter of fact, it can be shown that if the latent probabilities change through time, and if the manifest marginals are all the same, the marginals have to be equal to 0.50; but we shall not prove this here.

In the case of 0.50 constant marginals, we cannot simply apply to each successive pair of waves the formulas given above for the case of two waves, because, as we mentioned there, the formulas break down when the marginals are 0.50. We need to develop solutions which do not require (as in (5.10)) dividing by $(2p_i - 1)$, which is zero. To do this we can use some of the earlier equations derived for the two-wave model.

The structure of the present model is this:

Time	*Latent probability*	*Proportion in latent class 1*	*Proportion in latent class 2*
1	a_1	v_{111}	v_{222}
2	a_2	v_{111}	v_{222}
3	a_3	v_{111}	v_{222}

We have, from above, (5.6)

$$[12] = (2a_1 - 1)(2a_2 - 1)v_{111}(1 - v_{111})$$

Similarly

$$[13] = (2a_1 - 1)(2a_3 - 1)v_{111}(1 - v_{111}) \qquad (5.15)$$

$$[23] = (2a_2 - 1)(2a_3 - 1)v_{111}(1 - v_{111}) \qquad (5.16)$$

From these three equations, we get

$$\frac{[12][23]}{[13]} = (2a_2 - 1)^2 v_{111}(1 - v_{111})$$

so that

$$v_{111}(1 - v_{111}) = \frac{[12]\ [23]}{[13](2a_2 - 1)^2} \qquad (5.17)$$

From (5.8),

$$v_{111} = \frac{a_2 + p_2 - 1}{2a_2 - 1}$$

and

$$1 - v_{111} = \frac{a_2 - p_2}{2a_2 - 1}$$

Hence

$$v_{111}(1 - v_{111}) = \frac{(a_2 + p_2 - 1)(a_2 - p_2)}{(2a_2 - 1)^2} \qquad (5.18)$$

Combining (5.17) and (5.18) and eliminating $(2a_2 - 1)^2$,

$$\frac{[12]\ [23]}{[13]} = (a_2 + p_2 - 1)(a_2 - p_2) \qquad (5.19)$$

This can be solved for a_2, giving

$$a_2 = 0.5 \pm \sqrt{0.25 - p_2(1-p_2) + \frac{[12]\,[23]}{[13]}} \qquad (5.20)$$

Similarly, it can be shown that

$$a_1 = 0.5 \pm \sqrt{0.25 - p_1(1-p_1) + \frac{[12]\,[13]}{[23]}} \qquad (5.21)$$

and

$$a_3 = 0.5 \pm \sqrt{0.25 - p_3(1-p_3) + \frac{[13]\,[23]}{[12]}} \qquad (5.22)$$

To get v_{111}, we can use the three equations of the type of (5.8), taking an average for the three waves, as follows:

$$v_{111} = \frac{1}{3}\left(\frac{a_1 + p_1 - 1}{2a_1 - 1} + \frac{a_2 + p_2 - 1}{2a_2 - 1} + \frac{a_3 + p_3 - 1}{2a_3 - 1}\right) \qquad (5.23)$$

Example of the Case of Three Waves, 0.50 Marginals, Latent Probabilities Unrestricted

We shall use the data of the previous example, on grades of medical students. Since the data were given above, we need not show them again. The computations yield the following structure:

Time	*Latent probability*	*Proportion in latent class 1*	*Proportion in latent class 2*
1	0.760	0.520	0.480
2	0.871	0.520	0.480
3	0.846	0.520	0.480

We see that the latent probability jumps sharply between waves 1 and 2 (the first and second years of medical school), and declines by a fraction the third year. We can use this structure to generate the third order frequencies, as shown in Table 14.

This fit is so good that it creates the uneasy feeling that there may be no freedom in the system, but recall that the third order frequencies were not used for the solution. This model explains the data about as well as we can ever expect any model to explain them. Unfortunately, in the order of exposition, this example had to come near the beginning; after this all the other examples we can show may seem anti-climactic.

TABLE 14

Comparison of observed and estimated data on grades of medical students, unrestricted probabilities model

Grade position at			Estimated frequency	Observed frequency	Deviation
Time 1	*Time 2*	*Time 3*			
High	High	High	45.8	45	—0.8
High	High	Low	10.2	11	0.8
High	Low	High	9.1	10	0.9
High	Low	Low	14.5	14	—0.5
Low	High	High	15.5	16	0.5
Low	High	Low	8.8	8	—0.8
Low	Low	High	9.7	9	—0.7
Low	Low	Low	42.4	43	0.6
		Total	156.0	156	0.0

Three Waves, Changing Marginals, Unrestricted Probabilities

If the marginal proportions are not equal to 0.50, they will tend to change as the latent probabilities change. We shall have more to say about this later, but meanwhile we can use the solutions of the two wave model to solve the three wave model. The two wave model can be applied to every pair of the three waves, giving two solutions for each latent probability. From (5.11) and (5.12) we have

$$a_1 = 0.5 \pm \sqrt{0.25 - p_1(1-p_1) + [12]\frac{(2p_1-1)}{(2p_2-1)}}$$

$$a_2 = 0.5 \pm \sqrt{0.25 - p_2(1-p_2) + [12]\frac{(2p_2-1)}{(2p_1-1)}}$$

(It should be noted that in all these solutions with double roots, the larger value is the chosen latent probability, since a latent probability of less than 0.50 for two corresponding latent and manifest classes has no meaning in terms of these models.)

We can derive in exactly the same manner,

$$a_1 = 0.5 \pm \sqrt{0.25 - p_1(1-p_1) + [13]\frac{(2p_1-1)}{(2p_3-1)}} \qquad (5.24)$$

$$a_3 = 0.5 \pm \sqrt{0.25 - p_3(1-p_3) + [13]\,\frac{(2p_3-1)}{(2p_1-1)}} \qquad (5.25)$$

$$a_2 = 0.5 \pm \sqrt{0.25 - p_2(1-p_2) + [23]\,\frac{(2p_2-1)}{(2p_3-1)}} \qquad (5.26)$$

$$a_3 = 0.5 \pm \sqrt{0.25 - p_3(1-p_3) + [23]\,\frac{(2p_3-1)}{(2p_2-1)}} \qquad (5.27)$$

We thus have two solutions for each latent probability. By setting these equal to each other in pairs, it can be derived that the conditions which must hold for the data if the model is to fit are

$$[12](2p_3-1) = [13](2p_2-1) = [23](2p_1-1) \qquad (5.28)$$

Of course, with fallible data these equalities will not hold exactly. In practice, if the equalities are considered to be approximated closely enough to warrant the use of this model, the best procedure would be to estimate any given a_i by taking the arithmetic mean of the two estimates which can be obtained by the equations above. v_{111} can then be obtained by equation (5.23).

Example of Three Waves, Changing Marginals, Unrestricted Probabilities

For this model, we shall use a set of data which prima facie do not seem to fit very well. The variable is "Vote Intention" taken

TABLE 15

Vote intention in 1948, Elmira Panel, pairs of waves

	June				June				August		
	Rep.	Not Rep.			Rep.	Not Rep.			Rep.	Not Rep.	
Rep. Aug.	317	56	373	Rep. Oct.	320	55	375	Rep. Oct.	353	22	375
Not Rep.	45	144	189	Not Rep.	42	145	187	Not Rep.	20	167	187
	362	200	562		362	200	562		373	189	562

from the 1948 panel in Elmira, New York, which was used above in the two wave model. There we used only data from the August and October waves; here we use all three pre-election waves, taken in June, August, and October. The data for each pair of waves are shown in Table 15.

The reason why the model does not seem too promising for these data is that there appears to be a trend towards Republican between June and August, as evidenced by the turnover of 56 persons towards Republican as against only 45 who changed towards the Democratic or undecided positions (undecided being included with Democratic for this dichotomization). However, the manifest data can be very misleading in setting up these models, as will shortly appear.

First inspecting the data to see whether the conditions are met, we find that

$$[12](2p_3 - 1) = 0.0457$$

$$[13](2p_2 - 1) = 0.0457$$

$$[23](2p_1 - 1) = 0.0534$$

In the absence of a statistical theory to determine what difference between these supposedly equal numbers is significantly greater than zero, we may tentatively accept that the conditions are met. We then go on to use the formulas to compute values of a_i, obtaining values from each pair of waves separately, as shown in Table 16.

We observe that the two estimates of a_1 are almost the same, but the estimates of a_2 and a_3 from waves 2 and 3 are somewhat higher than those using the other pairs of waves — 0.956 and 0.966 as against 0.927 and 0.936, respectively. Whether these dif-

TABLE 16

Estimates of latent probabilities from different pairs of waves, and averages of these

Estimated from	*Estimated probabilities*		
	a_1	a_2	a_3
Waves 1 and 2	0.876	0.927	—
Waves 1 and 3	0.874	—	0.936
Waves 2 and 3	—	0.956	0.966
Average of all estimates	0.875	0.941	0.951

ferences are significant we again cannot presently say, but they seem large for a sample of size 562. However, we can decide the acceptability of this model better by recreating the data from it. For this purpose we shall use the averages of the different estimates of a_i, giving the following structure:

Time	*Latent probability*	*Proportion in latent class 1*	*Proportion in latent class 2*
1	0.875	0.688	0.312
2	0.941	0.688	0.312
3	0.951	0.688	0.312

It is noticeable that the latent probabilities go up sharply from wave 1 to wave 2 and slightly thereafter; in fact, in each of the separate estimates of the probabilities from pairs of waves we find them rising through time.

We go on to estimate all the third order frequencies from the model, using equations of the following sort:

$$p_{123} = a_1 a_2 a_3 v_{111} + (1-a_1)(1-a_2)(1-a_3)v_{222}$$

The estimates are shown in Table 17, together with the actual data.

We can see that the estimates are quite good for those who were "Not Republican" on the first wave, in June. However, among those who were "Republican" in June there are large deviations, centered around an underestimate for the group who moved to

TABLE 17

Estimated and observed third order frequencies, vote intention, 1948, Elmira Panel

Intention at			*Estimated frequency*	*Observed frequency*	*Deviation*
June	*August*	*October*			
Rep.	Rep.	Rep.	302.8	307	4.2
Rep.	Rep.	Not R.	16.8	10	−6.8
Rep.	Not R.	Rep.	19.9	13	−6.9
Rep.	Not R.	Not R.	20.6	32	11.4
Not R.	Rep.	Rep.	43.7	46	2.3
Not R.	Rep.	Not R.	10.8	10	−0.8
Not R.	Not R.	Rep.	9.8	9	−0.8
Not R.	Not R.	Not R.	137.6	135	−2.6
		Total	562.0	562	0.0

"Not Republican" in August and stayed in that manifest position in October. This underestimate, deviating by 11.4 from the observed value, actually forces two of the other estimates to be off (about 6.8 each), since certain sums must add up to equal the manifest data.

What is interesting about the fit is that this model has underestimated the number of those who shift from Republican in June to Not Republican in August, whereas the manifest data shows a trend towards Republican from June to August. The model has underestimated those moving in the opposite direction from the trend. We shall have more to say about this later, when another model is fitted to the same data. Meanwhile, the present model seems inadequate for the Elmira Vote Intention data.

CHAPTER 6

Models Involving both Latent Change and Change in Latent Probabilities

In order to identify latent process models we must limit the number of parameters of the models, i.e., the number of unknowns in the equations. In Chapter 3 we achieved this by putting a very severe restriction on the latent probabilities, namely, that they could not change at all. We were then able to solve the equations with practically no conditions on the movement among the latent classes. In the models of Chapter 5, we placed some restriction on the latent probabilities — in particular, that the probability of positive response for one latent class is the complement of the probability of positive response for the other — but we held the number of parameters down mainly by a harsh limitation on the latent positions, not permitting anyone to change his latent position.

However, there is good reason to think that latent probabilities change, in most cases, and there is also good reason to think that some people change their latent positions. By "mixed models" we refer to models in which some change is allowed in the latent probabilities and some change in the latent position, but henceforth we are going to consider chiefly models in which both are restricted sufficiently to permit a solution, with at least one degree of freedom. Let us consider the possibilities for a manifest dichotomy available at three time points. The number of independent items of information is, in general, seven ($2^3 - 1$). Sometimes it is fewer than seven because the model assumes certain equalities, but let us consider the case where there are seven independent proportions. If we want to leave any degrees of freedom, we must limit the model to six unknowns at most. Thus if we want to let the (complementary) latent probabilities change freely through time, there cannot be more than four latent classes (four, not three, since the latent proportions must add up to one). For example, we could permit a latent trend in one direction over both time intervals (without assuming linearity of the trend). Thus we would have the latent proportion of those who remain in latent class 1 all three times, v_{111}; the latent proportion of those who move from class 1 at time 2 to class 2 at time 3, v_{112}; the latent proportion of those who move from class 1 at time 1 to class 2 at time 2 v_{122}; and the latent proportion of those who are in class 2 at time 1 and therefore at times 2 and 3 (since the model does not permit movement out of latent class 2), v_{222}.

Suppose we want to permit latent movement in one direction between waves 1 and 2 and latent movement in the opposite direction between waves 2 and 3. Then it turns out that we need the following five latent classes:

v_{111}

v_{121}

v_{122}

v_{221}

v_{222}

This allows movement from latent class 1 at time 1 to latent class 2 at time 2, and from latent class 2 at time 2 to latent class 1 at time 3. Since there are four independent unknowns here, the number of independent latent probabilities must be limited to two if the equations are to be solved with any degrees of freedom. We must impose a restriction such as the equality of two of the latent probabilities, or linearity of the movement of the probabilities through time.

We cannot exhaust all possibilities in this volume. We are aided in selecting models at present by the fact that we have a purpose beyond the discussion of the models themselves. We want to show the applicability of the models to data. We have therefore chosen to discuss only those models which have relevance to the primary bodies of panel data referred to in the text. Two of the studies, both of which have been referred to previously, are studies of election campaigns, one of the 1940 campaign in Erie County, Ohio, one of the 1948 campaign in Elmira, New York. Both of these studies have served as the basis for volumes of considerable importance in the field of political sociology and social psychology. By using the data from these studies, we are assured that the phenomena we are trying to describe have substantive importance, so that if the models fit these data we can be assured they are not mere busy work. Other data are also used subsequently.

One Change in Latent Probability, One Change in Latent Position

We find some sets of data in which the amount of turnover between time 1 and time 2 is about the same as that between time 1 and time 3, while the turnover is smaller between time 2 and time 3. We have already looked at two such items, the grades of medical students and vote intention in the 1948 Elmira panel. In some cases, unlike those two, there is also a fairly sharp marginal or net shift between times 1 and 2 or between times 2 and 3, but not both. Since we have data which can be regenerated fairly well in terms of a latent change between time 2 and time 3, we shall describe that model here. We have the following latent structure:

Time	*Latent probability*	*Proportion in latent class 1*	*Proportion in latent class 2*
1	a_1	v_{111}	$v_{221} + v_{222}$
2	a_2	v_{111}	$v_{221} + v_{222}$
3	$a_3 = a_2$	$v_{111} + v_{221}$	v_{222}

In this model, the latent probability changes from time 1 to time 2, but remains constant from time 2 to time 3. The proportions in latent class 1 and 2, on the other hand, remain constant from time 1 to time 2, but change from time 2 to time 3, creating a third joint latent class over time, with the proportion v_{221} of those who move from latent class 2 at times 1 and 2 to latent class 1 at time 3.

There are several ways to solve the equations which result from this structure, but perhaps the easiest is to make use of the fact that the structure from time 2 to time 3 is exactly the same as that of the first model in Chapter 3, where latent change in one direction only was permitted without change in the latent probability. Equation (3.5) is the solution for the latent probability. Adapting the notation, this is

$$a_2 = a_3 = 0.5(1 + p_3 - p_2 + \sqrt{(1-p_2-p_3)^2 + 4[23]}) \tag{6.1}$$

We also have, from the previous case,

$$v_{111} = \frac{p_2 + a_2 - 1}{2a_2 - 1} \tag{6.2}$$

$$v_{111} + v_{221} = \frac{p_3 + a_2 - 1}{2a_2 - 1}$$

$$v_{221} = \frac{p_3 - p_2}{2a_2 - 1} \tag{6.3}$$

$$v_{222} = 1 - v_{111} - v_{221} \tag{6.4}$$

The only remaining unknown is a_1. This can be readily found by the method used in the model with no latent change but with latent probability changing from time 1 to time 2. The solution given in (5.13), modified slightly to fit the present case, is

$$a_1 = \frac{p_{12} - (1-a_2)(v_{221} + v_{222})}{a_2 v_{111} - (1-a_2)(v_{221} + v_{222})} \tag{6.5}$$

First Example of the Model: Intensity of Feeling about Political Choice

This example is taken from the Elmira panel. The data are based on the responses in June, August, and October to the question "Right now, how strongly do you feel about your choice — very strongly, only fairly strongly, or not very strongly at all?" The only difference in the item among the waves is that in June the question referred to party choice (since the candidates were not yet nominated) and in August and October the question referred to candidate choice. The answer categories are "Very strongly," "Fairly strongly," and "Not strongly." For our purpose the last two categories have been combined to yield a dichotomy which we shall call High and Low.

Taking the waves in pairs, the figures shown in Table 18 were obtained.

The figures show a substantial shift from Low to High between June and August and between August and October. They also

TABLE 18

Strength of feeling about political choice, Elmira Panel, pairs of waves

	June				*June*				*August*		
	High	Low			High	Low			High	Low	
High	159	81	240	High	174	101	275	High	209	66	275
Aug.				Oct.				Oct.			
Low	56	112	168	Low	41	92	133	Low	31	102	133
	215	193	408		215	193	408		240	168	408

show that the amount of turnover from June to August (137 cases) is about the same as from June to October (142 cases). The latter fact encourages us to apply a model in which the latent probabilities are the same at August and October, but the former fact may make it seem highly improbable that a model allowing latent change only between August and October can fit. It is important, therefore, to keep in mind that the movement among latent positions may be quite different from the movement among manifest positions, and not to prejudge the question. Let us apply this model and see what happens.

Using the equations given above, we obtain the following solutions for the latent structure:

Time	*Latent probability*	*Proportion in latent class 1*	*Proportion in latent class 2*
1	0.686	0.608	0.392
2	0.908	0.608	0.392
3	0.908	0.713	0.287

The latent probability makes a sharp jump up from 0.686 in June to 0.908 in August, and is by assumption the same in October. The proportion in the latent class whose strength of feeling about their choice is high is 0.608 in June and August (the same by assumption) and goes up to 0.713 in October.

The equations which we use to generate estimates of the third order frequencies are a little more complicated than those for the models which assumed no latent change. They are of the following type:

$$p_{123} = a_1 a_2 a_3 v_{111} + (1-a_1)(1-a_2)a_3 v_{221} + (1-a_1)(1-a_2)(1-a_3)v_{222}$$

It is necessary to include an extra term to take care of the latent proportion v_{221}. Using equations of this type, the estimated third order frequencies were computed and are shown below.

We can see that this model is not a bad fit at all, despite the fact that the assumptions seemed a little at variance with the manifest situation. The fact is that the shift in latent probability from June to August caused the estimated marginals to shift, even though no latent change was admitted. The estimated and observed figures for the relation between June and August alone are as follows:

		Estimated			*Observed*	
	June	High	Low		High	Low
August	High	158.8	80.9	High	159	81
	Low	61.2	107.1	Low	56	112

The change in latent probability (Table 19) yielded a net shift of about 20 towards High, as against an observed net shift of 25. The discussion of such marginal shifts in the manifest data without any corresponding latent shift will be deferred. At the moment, we may observe that the deficiency of 5 in the shift yielded by the model means that we could obtain a better fit by allowing true change in the same direction between June and August. We could then obtain a still better fit by allowing the latent probability to change between August and October as well as between June and

TABLE 19

Estimated and observed third order frequencies, strength of feeling about political choice, 1948, Elmira Panel

Strength of feeling at			*Estimated frequency*	*Observed frequency*	*Deviation*
June	*August*	*October*			
High	High	High	141.7	146	4.3
High	High	Low	17.1	13	—4.1
High	Low	High	28.2	28	—0.2
High	Low	Low	33.0	28	—5.0
Low	High	High	66.2	63	—3.2
Low	High	Low	14.7	18	3.3
Low	High	High	38.9	38	—0.9
Low	Low	Low	68.2	74	5.8
			408.0	408	0.0

August. However, the present model fits closely enough to show that models of this type can handle the data reasonably well, and we shall not deal further with this item.

Second Example of the Model: Main Source of Information about Political Campaign

The data are taken from the study of the 1940 political campaign in Erie County, Ohio. They differ from every other example in this book in that the question permitted more than one response. The question was "Where do you think you will get most of your information about issues and candidates in the coming presidential elections: from talking to friends or relatives, from seeing newsreels, from going to political meetings, from magazines, radio, newspapers, or where?" We have selected two of the responses, "radio" and "newspapers", and eliminated the others, for present purposes. The only respondents included here are those who answered either "radio" or "newspapers" or both at each of three waves, in May, August, and October. The unit of analysis, however, is the response pattern rather than the respondent, since multiple responses are treated as separate responses.

The data taking the three waves pairwise are shown in the table below.

TABLE 20

Main source of information about political campaign, 1940, Erie County Panel, Pairs of waves

	May				*May*				*August*		
	Rad.	News.			Rad.	News.			Rad.	News.	
Rad.	156	114	270	Rad.	162	122	284	Rad.	168	116	284
Aug.				Oct.				Oct.			
News.	117	174	290	News.	111	166	277	News.	102	175	277
	273	288	561		273	288	561		270	291	561

We observe the familiar phenomenon that the turnover between the first and third waves (233 cases) is about the same as the turnover between the first and second waves (231 cases), while the turnover is smaller between the second and third waves (218 cases). (The frequency of this phenomenon in the examples cited should not mislead the reader into thinking that it always occurs. We have seen in previous chapters and will see presently cases where it does not occur. The fact that it does not occur, however, does not mean that models of this general type will not fit the data.)

The table also shows that there is almost no marginal shift from May to August, but from August to October there is a net shift of 14 cases towards "radio". The model used for the previous example therefore seems reasonable. Computations of the structure yield the following values for the parameters:

Time	*Latent probability*	*Proportion in latent class 1*	*Proportion in latent class 2*
1	0.678	0.462	0.538
2	0.749	0.462	0.538
3	0.749	0.512	0.488

The latent probability goes up from 0.678 in June to 0.749 in August and October. The fact that these probabilities are so low shows a large random element for this item, possibly a result in part of the fact that the item has low reliability, and partly the result of the multiple responses. Five per cent of the patterns show a shift from the latent class of "newspaper" to the latent class of "radio". Apparently radio assumed greater importance as the campaign wore on.

TABLE 21

Estimated and observed third order frequencies, main source of information about political campaign, 1940, Erie County Panel

Main source of information at			*Estimated frequency*	*Observed frequency*	*Deviation*
May	*August*	*October*			
Radio	Radio	Radio	106.0	106	0.0
Radio	Radio	Newsp.	50.0	50	0.0
Radio	Newsp.	Radio	54.4	56	1.6
Radio	Newsp.	Newsp.	62.8	61	−1.8
Newsp.	Radio	Radio	62.3	62	−0.3
Newsp.	Radio	Newsp.	51.5	52	0.5
Newsp.	Newsp.	Radio	61.2	60	−1.2
Newsp.	Newsp.	Newsp.	112.8	114	1.2
			561.0	561	0.0

The estimates of third order frequencies generated from the model are shown in Table 21.

The fit is almost perfect, but the use of response patterns instead of respondents as the unit of analysis partly biases the data in our favor. Any instance where a person responds both "radio" and "newspaper" on the same wave will bring about parallel patterns of response which will fit the pattern of randomness.

One Change in Latent Probability, Change in Latent Position in Same Direction over each Interval

Here we elaborate the previous model in only one way, by allowing latent change between time 1 and time 2 in the same direction in which it occurs between time 2 and time 3. Thus we postulate latent trend over the entire time, but we do not require that it be linear. The model thus has the following structure:

Time	*Latent probability*	*Proportion in latent class 1*	*Proportion in latent class 2*
1	a_1	$v_{111} + v_{112} + v_{122}$	v_{222}
2	a_2	$v_{111} + v_{112}$	$v_{112} + v_{222}$
3	$a_3 = a_2$	v_{111}	$v_{112} + v_{122} + v_{222}$

The situation for the second and third waves is exactly the same as that for the previous model (except that the notation is a little different), and the same method can be used to solve for a_2, v_{111}, v_{112}, and the sum $(v_{122} + v_{222})$. We need to use the equations (6.1)—(6.4) with appropriate changes of notation and subscripts.

We then have, from the fundamental assumptions,

$$p_1 = a_1(v_{111} + v_{112}) + a_1 v_{122} + (1-a_1)v_{222} \tag{6.6}$$

Multiplying both sides by $(1-a_2)$,

$$(1-a_2)p_1 = a_1(1-a_2)(v_{111} + v_{112}) + a_1(1-a_2)v_{122} + (1-a_1)(1-a_2)v_{222} \tag{6.7}$$

Also from the fundamental assumptions,

$$p_{12} = a_1 a_2(v_{111} + v_{112}) + a_1(1-a_2)v_{122} + (1-a_1)(1-a_2)v_{222} \tag{6.8}$$

Subtracting (6.7) from (6.8) and solving for a_1,

$$a_1 = \frac{p_{12} - (1 - a_2)p_1}{(2a_2 - 1)(v_{111} + v_{112})} \tag{6.9}$$

Since we know v_{111}, v_{112}, and a_2, we can find the value of a_1.

From this we can get v_{221} and v_{222} by equations of the type of (6.2)—(6.4), such as

$$v_{111} + v_{112} + v_{122} = \frac{p_1 + a_1 - 1}{2a_1 - 1} = 1 - v_{222} \tag{6.10}$$

Example of the Model: Vote Intention in 1940, Erie County Panel

We have selected for the application of this model the data from the final three pre-election waves of the Erie County political study. The item is the vote intention of the respondent and the responses used here were taken in August, September, and October. We have dichotomized the responses, grouping "Roosevelt" with "Don't know" as against "Willkie" (because this gives a split nearest fifty per cent). However, we shall identify the "Not Willkie" group with latent class 1. The data are shown in Table 22.

TABLE 22

Vote intention, Erie County, 1940, in pairs of waves

	August				*August*				*September*		
	Not W.	Willkie			Not W.	Willkie			Not W.	Willkie	
Not W.	239	8	247	Not W.	230	10	240	Not W.	234	6	240
Sept.				Oct.				Oct.			
Willkie	14	184	198	Willkie	23	182	205	Willkie	13	192	205
	253	192	445		253	192	445		247	198	445

There is very little manifest change in the data. The amount of turnover from the first to the third waves (33 cases) is definitely greater than for the first and second waves (22) or for the second and third (19), which are rather similar. There is a steady trend towards Willkie, which warrants trying out the present model.

Solving by means of the equations above, the following values are obtained for the parameters of the structure:

Time	*Latent probability*	*Proportion in latent class 1*	*Proportion in latent class 2*
1	0.958	0.540 + 0.016 + 0.019 = 0.575	0.425
2	0.987	0.540 + 0.016 = 0.556	0.019 + 0.425 = 0.444
3	0.987	0.540	0.016 + 0.019 + 0.425 = 0.460

The latent probability increases from August to September. The latent trend turns out to be roughly linear, with 0.019 going towards Willkie from August to September and 0.016 going towards Willkie from September to October.

In comparing the third order frequencies generated by this model to the actual data, we have the difficulty that some of the observed frequencies are extremely small, because of the small amount of turnover. The comparison is given in Table 23.

TABLE 23

Estimated and observed third order frequencies, vote intention, Erie County Panel, 1940

Vote intention in			*Estimated frequency*	*Observed frequency*	*Deviation*
August	*September*	*October*			
Not W.	Not W.	Not W.	224.5	229	4.5
Not W.	Not W.	Willkie	9.8	10	0.2
Not W.	Willkie	Not W.	3.2	1	−2.2
Not W.	Willkie	Willkie	15.6	13	−2.6
Willkie	Not W.	Not W.	9.6	5	−4.6
Willkie	Not W.	Willkie	2.8	3	0.2
Willkie	Willkie	Not W.	2.5	5	2.5
Willkie	Willkie	Willkie	177.0	179	2.0
			445.0	445	0.0

This fit would not be very good in terms of chi-square, but the frequencies are too small for such a test anyway. The fit is actually fairly good except for the deviation of − 4.6 for those moving away from Willkie. It is more instructive in this particular case to look at the generated data in comparison to the observed for the relation between wave 1 and wave 3, keeping in mind that the

TABLE 24

Estimated and observed joint frequencies for August and October, vote intention, Erie County Panel, 1940

		Estimated				*Observed*	
		August				August	
		Not W.	Willkie			Not W.	Willkie
October	Not. W.	227.7	12.1	October	Not W.	230	10
	Willkie	25.4	179.8		Willkie	23	182

joint frequency p_{13} was not used in computing the parameters of the structure. The comparison is shown in Table 24.

In this comparison we can see that the model is a fairly good fit. The combination of a latent trend with an increase in latent probabilities predicts the data reasonably well when larger numbers are used. It would be possible to improve the fit by complicating the model a little more, of course.

One Change in Latent Probability, Latent Change in One Direction Over First Interval and in Opposite Direction over Second Interval

The model here is similar to the previous one, with the exception that latent change from time 1 to time 2 is postulated in the opposite direction rather than the same direction of the latent change from time 2 to time 3. The structure of the model then looks like this:

Time	*Latent probability*	*Proportion in latent class 1*	*Proportion in latent class 2*
1	a_1	$v_{111} + v_{112}$	$v_{211} + v_{212} + v_{222}$
2	a_2	$v_{111} + v_{112} + v_{211} + v_{212}$	v_{222}
3	$a_3 = a_2$	$v_{111} + v_{211}$	$v_{112} + v_{212} + v_{222}$

The solution has been obtained but will not be presented here. Instead, we shall briefly describe an example for which the model provides a moderately good fit. The item is taken from the Elmira panel and reads "How much interest would you say you have in this year's presidential election — a great deal, quite a lot, not very much, or none at all?" 754 respondents answered this question on all three waves. The responses have been dichotomized into "a great deal" and "other" for the example. The data show a considerable increase in interest from June to August, and a slight decline from August to October. When the structure is computed, the following values for the parameters are obtained:

Time	*Latent probability*	*Proportion in latent class 1*	*Proportion in latent class 2*
1	0.821	0.211 + 0.017 = 0.228	0.069 + 0.003 + 0.700 = 0.772
2	0.843	0.211 + 0.017 + 0.069 + 0.003 = 0.300	0.700
3	0.843	0.211 + 0.069 = 0.280	0.017 + 0.003 + 0.700 = 0.720

When the third order frequencies are generated from the model and compared with the observed data, the figures in Table 25 below are obtained.

TABLE 25

Estimated and observed third order frequencies, interest in election, Elmira Panel, 1948

Interest in			*Estimated frequency*	*Observed frequency*	*Deviation*
June	*August*	*October*			
High	High	High	103.1	97	−6.1
High	High	Low	39.1	45	5.9
High	Low	High	31.6	38	6.4
High	Low	Low	71.4	65	−6.4
Low	High	High	61.7	68	6.3
Low	High	Low	69.9	64	−5.9
Low	Low	High	66.9	60	−6.9
Low	Low	Low	310.3	317	6.7
			754.0	754	0.0

The fit is only fair, and it is likely that a better fit would be obtained by eliminating the latent change from time 2 to time 3, and instead allowing the latent probabilities to change from time 2 to time 3, which would still leave one degree of freedom.

Latent Probabilities Free To Change, One Latent Change

In models of this sort, we permit the latent probability (always assuming the probability for one latent class to be the complement of that of the other) to change freely over each time interval. We also postulate that there will be a change in the latent position in one direction over one of the time intervals. This makes five independent unknowns. If we assume latent change from latent class 1 to latent class 2 over the first interval, the structure looks like this:

Time	*Latent probability*	*Proportion in latent class 1*	*Proportion in latent class 2*
1	a_1	$v_{111} + v_{122}$	v_{222}
2	a_2	v_{111}	$v_{122} + v_{222}$
3	a_3	v_{111}	$v_{122} + v_{222}$

We can readily solve the resulting equations by a combination of methods already shown, and we shall not repeat these solutions. The unknowns for waves 2 and 3 can be obtained by the solution for the first model of Chapter 5, a model for two waves with latent probability free to change but no latent change. Once these solutions are obtained, the unknowns for wave 1 can be obtained by the method used in the mixed model with one change in latent probability and a latent shift in the same direction over both time intervals.

Example of the Model: Vote Intention in the Elmira Panel

We come now to one of the most interesting and paradoxical findings of this volume. We attempted above to fit a model involving only changes in the latent probabilities to the vote intention data from the Elmira panel. We found that the fit was generally good except for one cell, where the model underestimated the number of people shifting from Republican between June and August. Accordingly, we shall now use a model in which latent change away from Republican is postulated between June and August, even though the manifest data shows a trend towards Republican during that time. (Class 1 is Republican.)

When the parameters are computed by the method suggested above, the following values are obtained:

Time	*Latent probability*	*Proportion in latent class 1*	*Proportion in latent class 2*
1	0.857	0.702	0.298
2	0.956	0.686	0.314
3	0.966	0.686	0.314

We see that the latent probability rises sharply from June to August, with a further slight rise by October. The proportion of latent Republicans in June is 0.702, and this proportion decreases to 0.686 by August, even though the manifest proportion of Republicans increases.

When the third order frequencies are estimated by the model we obtain the following results (Table 26).

TABLE 26

Estimated and observed third order frequencies, vote intention, Elmira Panel, 1948

Vote intention in			*Estimated frequency*	*Observed frequency*	*Deviation*
June	*August*	*October*			
Rep.	Rep.	Rep.	305.0	307	2.0
Rep.	Rep.	Not R.	12.0	10	−2.0
Rep.	Not R.	Rep.	15.0	13	−2.0
Rep.	Not R.	Not R.	29.9	32	2.1
Not R.	Rep.	Rep.	51.2	46	−5.2
Not R.	Rep.	Not R.	7.9	10	2.1
Not R.	Not R.	Rep.	7.0	9	2.0
Not R.	Not R.	Not R.	133.9	135	1.1
			561.9	562	0.1 (Rounding error)

The model is now an excellent fit, except that we have erred slightly in the opposite direction of overestimating by 5.2 the number who shifted towards Republican from June to August.

The fact that the model fits confirms a point of considerable interest, namely, that with changing latent probabilities the latent and observed shifts may be in opposite directions. Between June and August, the manifest data showed a two per cent net shift *towards* Republican, while the latent shift was 0.016 *away* from the Republicans. This is particularly interesting because of our knowledge after the fact that, while everyone thought the Republicans were going to win, Truman actually won. The manifest shift toward Republican is produced by the increase in latent probability, so that respondents are moving towards their latent positions. This movement is strong enough to override a latent shift away from Republican in terms of the appearance of the manifest data.

Another Example of the Model: Winner Expectations in the Elmira Panel

In Chapters 2, 3, 5 and 6 up to this point we have examined every item available for three waves from the Elmira and Erie County studies, with one exception, an item in the Elmira study reading as follows: "Regardless of which candidate you yourself might vote for, which one do you think is actually going to win the presidential election?" (In June, read "party" instead of "candidate".) For the sake of comprehensiveness, we now look at this item.

The data in pairs of waves are as follows (Table 27):

TABLE 27

Winner expectations, 1948, Elmira Panel, in pairs of waves

	June				*June*				*August*		
	Rep.	Not Rep.			Rep.	Not Rep.			Rep.	Not Rep.	
Rep. Aug.	328	80	408	Rep. Oct.	326	77	403	Rep. Oct.	362	41	403
Not Rep.	39	61	100	Not Rep.	41	64	105	Not Rep.	46	59	105
	367	141	508		367	141	508		408	100	508

The table shows that the amount of turnover from June to August (119 cases) is about the same as from June to October (118 cases) and greater than the amount from August to October (87 cases). Since the marginals do not shift much between August and October, the model used in the previous example seems appropriate.

The computed values for the structure are as follows:

Time	*Latent probability*	*Proportion in latent class 1*	*Proportion in latent class 2*
1	0.878	0.794	0.206
2	0.912	0.868	0.132
3	0.899	0.868	0.132

When the third order frequencies are generated from the model, the following estimates result:

TABLE 28

Estimated and observed third order frequencies, winner expectations, Elmira Panel, 1948

June	*August*	*October*	*Estimated frequency*	*Observed frequency*	*Deviation*
Rep.	Rep.	Rep.	294.3	302	7.7
Rep.	Rep.	Not R.	33.7	26	−7.7
Rep.	Not R.	Rep.	29.1	24	−5.1
Rep.	Not R.	Not R.	9.9	15	5.1
Not R.	Rep.	Rep.	67.8	60	−7.8
Not R.	Rep.	Not R.	12.2	20	7.8
Not R.	Not R.	Rep.	11.9	17	5.1
Not R.	Not R.	Not R.	49.1	44	−5.1
			508.0	508	0.0

We can see that the fit is rather poor, and that this model is not suitable for these data.

Latent Probabilities Unrestricted, Latent Change in Same Direction Over Two Time Periods

It is often the case that there is reason to assume change on the latent level only in one direction, frequently because intervening events or stimuli (which may be accidental or deliberately brought about) are supposed on a common-sense or theoretical basis of some sort to push people in a certain direction. In such situations a reasonable structure would look as follows:

Time	*Latent probability*	*Proportion in latent class 1*	*Proportion in latent class 2*
1	a_1	$v_{111} + v_{112} + v_{122}$	v_{222}
2	a_2	$v_{111} + v_{112}$	$v_{122} + v_{222}$
3	a_2	v_{111}	$v_{112} + v_{122} + v_{222}$

If, as before, we let

$$[p_{12}] = p_{11.} - p_{1..}p_{.1.}$$

that is, the cross-product of time 1 and time 2, and so on, and we now define quantities A, B, and C as follows:

$$A = [p_{23}]$$

$$B = p_{.1.}(2p_{1.1} - p_{1..}p_{..1}) - p_{1..}(2p_{.11} - p_{.1.}p_{..1}) - (p_{1.1} + p_{1..}p_{..1})$$

$$C = p_{1..}[p_{13}] - p_{11.}[p_{13}] - p_{1..}p_{.1.}p_{1.1} + p_{1..}^2 p_{.11}$$

we get the solutions

$$a_1 = \frac{-B + \sqrt{B^2 - 4AC}}{2A} \tag{6.11}$$

$$a_2 = 1 - \frac{p_{11.} - a_1 p_{.1.}}{p_{1..} - a_1} \tag{6.12}$$

$$a_3 = 1 - \frac{p_{1.1} - a_1 p_{..1}}{p_{1..} - a_1} \tag{6.13}$$

We can then derive

$$v_{1..} = \frac{p_{1..} + a_1 - 1}{2a_1 - 1}$$

$$v_{.1.} = \frac{p_{.1.} + a_2 - 1}{2a_2 - 1}$$

$$v_{..1} = \frac{p_{..1} + a_3 - 1}{2a_3 - 1}$$

from which, in turn, we get

$$v_{111} = v_{..1} \tag{6.14}$$

$$v_{112} = v_{.1.} - v_{..1} \tag{6.15}$$

$$v_{122} = v_{1..} - v_{111} - v_{112} \tag{6.16}$$

$$v_{222} = 1 - v_{1..} \tag{6.17}$$

Examples of the Model

We have available a number of examples from two studies. The first study was an evaluation of an elaborate nation-wide program called by the acronym ENABLE (for Education and Neighborhood Action for Better Living Environment). The entire issue of the journal *Social Casework* for December, 1967 (Vol. XLVIII, No. 10) is devoted to a description of the action project, in a set of five articles, and the design and substance will not be detailed here. Briefly, it entailed the location in some 61 low-income communities across the nation of parents who engaged in group discussions primarily centered on problems of raising children, but also provided a variety of services including referrals and community action where needed or desired by the parents. The program was funded by the Office of Economic Opportunity. The teaching materials for group leaders and their assistants, who were usually local people previously untrained (sometimes referred to as "paraprofessionals" or "indigenous workers"), were prepared largely by the Child Study Association of America, which also had major responsibility for training the leaders. The community action aspects of the program were under the direction of the National Urban League. Major responsibility for the organization and administration of the program was given to the Family Service Association of America, which provided the overall project director, Mrs. Ellen Manser.

Just as these three national organizations worked closely together in administering the program, so did they work closely with the independent outside organization, the Simulmatics Corporation, which was responsible for evaluation of the project. The project had an inside research director, Dr. Aaron Rosenblatt, who provided among other things the necessary elaborate liaison of the project with the evaluation. The evaluation was directed by the present writer. The panel portion of the evaluation is not described in the journal referred to above, but is described in detail in a lengthy unpublished document entitled "Evaluation of Project ENABLE" (1967), available in the files of the Office of Economic Opportunity and the library of the Family Service Association of America.

Basically, the design of the panel consisted of a pre-program interview, and an interview with many identical items at the end of the program. The program typically lasted about eight or ten weeks. The program was run in two series, and it was possible for the first series to obtain a third follow-up interview at approximately the same time interval after the post-interview, for some 1,014 parents in all.

Substantively, the interesting feature of this study for the present models is the fact that the first interval constituted a quasi-experiment, in a rough way, while the second interval represented little or no intervention. Some of the repeated items had a predicted direction of change, while some did not, for the period of intervention. For the next interval, there was often no theoretical reason to think that there would be a further effect in the same direction, a mere maintenance of an effect already achieved, a "sleeper" effect, or a movement back to the pre-test position. For that reason, and because of the diversity of objectives and programs with different groups and individuals, a variety of subject matter was tapped. The evaluation report gives the complete details on the manifest level, using averages and including the third wave, but it does not include any attempts to examine latent structure models such as we are doing here.

Subsequently, for the purpose of the present work, a large number of the three-wave items were examined, and some of those which seemed to fit the present model are reported here. The response metrics of the various items differed in various ways, according to subject matter, and the subject matters often represented groupings of items similar in format.

The first item shown here is from a group of items asking about parental attitudes towards raising children, as well as self-reports about the attitudes and behavior of the parents towards their own children. The responses here were on a graphic thermometer with marks from zero to ten. For the present purpose, the responses have been dichotomized as near as possible to the median at time 1. The item is "I yell at my children more than I should." The

TABLE 29

Answers to question "I yell at my children more than I should", project ENABLE panel, in pairs of waves spaced eight weeks apart

	Time 1				Time 1				Time 2		
	High	Low			High	Low			High	Low	
High	340	130	470	High	324	134	458	High	318	140	458
Time 2				Time 3				Time 3			
Low	189	331	520	Low	205	327	532	Low	152	380	532
	529	461	990		529	461	990		470	520	990

supposition of the program directors was that, whatever the actual family situation, it was not generally likely to be beneficial for the parent to feel that he or she yelled at their children more than they themselves thought they should. The manifest data for pairs of waves are shown in Table 29.

The amount of turnover from the first to the second wave is 319 cases; from time 2 to time 3 it is 292 cases; and from time 1 to time 3 it is 339 cases. The number on the high end of the scale (that is, the parents who were more likely to think that they yelled at their children too much) declined from 529 at time 1 to 470 at time 2 to 458 at time 3. Thus there was relatively large decline during the period of intervention and a modest decline during the follow-up period. The model above seems reasonable, and the computed values for the structure are as follows:

Time	*Latent probability*	*Proportion in latent class 1*	*Proportion in latent class 2*
1	0.82	0.55	0.45
2	0.84	0.46	0.54
3	0.82	0.44	0.56

When the third order frequencies are generated from the model, the following estimates result (Table 30).

The model is a very close fit, but this seems more the result of allowing latent change, where the latent marginals decline from 0.55 to 0.44, than from freedom for the latent probabilities. Nevertheless, the latter do change slightly, and the point to be noted is

TABLE 30

Estimated and observed third order frequencies, responses to item "I yell at my children more than I should", ENABLE panel

Time 1	*Time 2*	*Time 3*	*Estimated frequency*	*Observed frequency*	*Discrepancy*
High	High	High	252.7	254	1.3
High	High	Low	87.3	86	−1.3
High	Low	High	71.3	70	−1.3
High	Low	Low	117.7	119	1.3
Low	High	High	65.3	64	−1.3
Low	High	Low	64.7	66	1.3
Low	Low	High	68.7	70	1.3
Low	Low	Low	262.3	261	−1.3
			990.0	990	0.0

that the highest latent probability, 0.84, occurs at time 2, the end of the intervention, in contrast to most of the previously studied data allowing latent probabilities to change, where the highest latent probability usually came at the end of the series. Here, time 2 is really the rough equivalent of the end of, say, a political campaign, and despite the continuation in the movement of marginals to a slight degree, there was less structurization of the situation at time 3 than at time 2. In the concluding chapter we shall discuss further some of the implications of this kind of finding in relation to the substantive situation involved, as well as theory bearing on the matter.

A second item chosen from the same study and using the same model is drawn from a group of items about aspirations of the parents for their children. Again, the item is a median dichotomization of a graphic rating scale running from zero to ten. It asks "How satisfied would you be if your child became an entertainer?". Here, because the marginals of high satisfaction increased over time, the order of responses has been reversed to make the response "low" in effect the positive response, for ease in computing the parameters. The pairwise data will not be given separately (they can be easily added from the third order data given), but it might be noted that the turnover from time 1 to time 2 was 299 cases; from time 2 to time 3 it was 257 cases, and from time 1 to time 3 it was 306 cases, only seven more cases than from time 1 to time 2. The computed values for the structure are as follows:

Time	*Latent probability*	*Proportion in latent class 1 (low)*	*Proportion in latent class 2*
1	0.78	0.46	0.54
2	0.82	0.39	0.61
3	0.85	0.34	0.66

When the third order frequencies are generated from the model, the following estimates result (Table 31).

Here the model is a close fit, although the latent probabilities do not quite conform to substantive expectations, going up from time 1 to time 2 but continuing to increase slightly from time 2 to time 3. This may be because the continuing decline in marginals resulted in a considerable departure from a fifty-fifty split, despite selection of the cut closest to the median, and that may have the effect of slightly elevating the latent probabilities at time 3. This point will be discussed in the next chapter.

A third item taken from the ENABLE panel is from a series

TABLE 31

Estimated and observed third order frequencies, responses to item "How satisfied would you be if your child became an entertainer?", ENABLE panel

Time 1	Time 2	Time 3	Estimated frequency	Observed frequency	Discrepancy
Low	Low	Low	167.3	166	−1.3
Low	Low	High	75.7	77	1.3
Low	High	Low	53.7	55	1.3
Low	Low	Low	117.3	116	−1.3
High	Low	High	56.7	58	1.3
High	Low	High	71.3	70	−1.3
High	High	Low	56.3	55	−1.3
High	High	High	265.7	267	1.3
			864.0	864	0.0

about knowledge of community services and resources available to low income parents. Here the question was "If a husband and wife are fighting a lot, where could they go to get help free or almost free to keep the marriage together?" The responses were coded into a dichotomy, in which the statement of a specific resource or service was one category, while all other responses fell into the other. For convenience in computing the model, the former was called "negative", although of course the aim was to improve knowledge of services. By so designating the responses, the proportion "positive" on the manifest level fell from 0.62 at time 1 to 0.46 at time 2 to 0.44 at time 3. In other words, knowledge of specific resources and services increased substantially during the period of intervention, and very slightly during the follow-up period.

The turnover from time 1 to time 2 was 380 cases; from time 2 to time 3 it was 277 cases; and from time 1 to time 3 it was 395 cases. The computed values for the structure are as follows:

Time	*Latent probability*	*Proportion in latent class 1*	*Proportion in latent class 2*
1	0.82	0.68	0.32
2	0.84	0.44	0.56
3	0.84	0.41	0.59

Estimates and observed values of the third order frequencies are as follows:

TABLE 32

Estimates and observed third order frequencies, responses to item on knowledge of community resources for marriage counselling, ENABLE panel

Time 1 knowledge	*Time 2 knowledge*	*Time 3 knowledge*	*Estimated frequency*	*Observed frequency*	*Discrepancy*
No	No	No	244.4	241	3.4
No	No	Yes	96.6	100	−3.4
No	Yes	No	77.6	81	−3.4
No	Yes	Yes	189.4	186	3.4
Yes	No	No	59.6	63	−3.4
Yes	No	Yes	53.4	50	3.4
Yes	Yes	No	49.4	46	3.4
Yes	Yes	Yes	214.6	218	−3.4
			985.0	985	0.0

In relation to the numbers involved, this is a reasonably good fit to the data (for those interested, chi-square is 1.07). The latent probability goes up from 0.82 to 0.84 between times 1 and 2, and remains at 0.84 between times 2 and 3, which accords fairly well with the substantive situation.

Another Example of the Same Model Using an Unchanging Manifest Stratifying Variable

If a sample is sufficiently large, it is possible to introduce an explicit stratifier to explore the possibility that a given model fits different categories of the group studied in the same or different ways. This is the case in Project ENABLE, and one more item from that study will be explored in terms of the same model as above but with the addition of a stratifying variable.

First we may look at the model as applied to the entire group studied, for which 979 responses for three waves were available. The item was as follows: "In the past week did any of your children come to you with a problem that was bothering them?" The responses were dichotomized into the categories "yes" and "no", and since the thrust of the program was to help increase communications between parents and children, the expected and actual movement of the marginals over time was towards an increase in the "yes" response. However, for simplicity of computation in terms of the model, the "yes" response was called negative and the "no" response positive. Thus the movement was a decrease in the proportion giving the positive response, from 0.57 at time 1 to 0.48 at time 2 to 0.42 at time 3. The turnover for the whole group was 342 cases between time 1 and time 2, 307 cases between time 2 and time 3, and 395 cases between time 1 and time 3, the kind of pattern we have seen so frequently. The computed values for the structure are as follows:

Time	*Latent probability*	*Proportion in latent class 1*	*Proportion in latent class 2*
1	0.80	0.62	0.38
2	0.86	0.47	0.53
3	0.83	0.37	0.63

Estimates and observed values of the third order frequencies are as follows:

TABLE 33

Estimated and observed third order frequencies, responses to item "In the past week did any of your children come to you with a problem that was bothering them?", ENABLE panel

Time 1	*Time 2*	*Time 3*	*Estimated frequency*	*Observed frequency*	*Discrepancy*
No	No	No	222.6	223	−0.4
No	No	Yes	122.4	122	0.4
No	Yes	No	64.4	64	0.4
No	Yes	Yes	152.6	153	−0.4
Yes	No	No	62.4	62	0.4
Yes	No	Yes	62.6	63	−0.4
Yes	Yes	No	57.6	58	−0.4
Yes	Yes	Yes	234.4	234	0.4
			979.0	979	0.0

The fit is excellent, and the structure shows considerable latent movement in the desired direction, continuing in fact through the interval between times 2 and 3. Further, the latent probabilities go up from time 1 to time 2 and fall from time 2 to time 3, thus conforming well to expectations for this type of situation.

In the ENABLE project, there was no attempt made to move different groups of parents' relations to their children or to their self-concept of their relations to their children in different ways in terms of the item we have been examining. Thus the movement of parents with greater education would not be expected to be any different from that of parents with less education, although that of course could happen.

The education of parents was associated, although to a rather low degree, with the responses to the item at time 1. Parents were dichotomized into the nearest possible median split on number of years of education. For four of the 979 parents used in the model above, education was not obtained. There were 490 placed in the "high education" group, here defined as going through the eleventh grade or more, and of these 0.55 gave the "no" response at time 1. 485 were placed in the "low education" group, here defined as having achieved ten or fewer grades of school, and of these 0.61 gave the "no" response at time 1.

In presenting the results of attempts to fit the model at hand to each of these classes, the estimates, observed values, and discrepancies will be given first for the low education group (Table 34) then for the high education group (Table 35). After that, the computed values of the models will be given in one table for comparison.

The manifest proportion giving the "no" response declined from 0.61 at time 1 ro 0.51 at time 2 to 0.47 at time 3. The turnover was 172 cases from time 1 to time 2, 157 cases from time 2 to time 3, and 193 cases from time 1 to time 3.

For the high income parents, the corresponding proportions giving the "no" response were 0.54 at time 1, 0.45 at time 2, and

TABLE 34

Estimates and observed third order frequencies, responses to item "In the past week did any of your children come to you with a problem that was bothering them?", low education parents, ENABLE panel

Time 1	*Time 2*	*Time 3*	*Estimated frequency*	*Observed frequency*	*Discrepancy*
No	No	No	125.7	129	–3.3
No	No	Yes	60.3	57	3.3
No	Yes	No	38.3	35	3.3
No	Yes	Yes	70.7	74	–3.3
Yes	No	No	33.3	30	3.3
Yes	No	Yes	29.7	33	–3.3
Yes	Yes	No	28.7	32	–3.3
Yes	Yes	Yes	98.3	95	3.3
			485.0	485	0.0

0.37 at time 3. The turnover was 170 cases from time 1 to time 2, 140 from time 2 to time 3, and 202 from time 1 to time 3. Thus the pattern of manifest data was very similar for the two education groups, though the initial position was somewhat different, and the follow-up movement of the better educated group was greater than for the less educated parents.

The estimated and observed third order frequencies for the better educated parents were as follows (Table 35):

TABLE 35

Estimated and observed third order frequencies, responses to item "In the past week did any of your children come to you with a problem that was bothering them?", high education parents, ENABLE panel

Time 1	*Time 2*	*Time 3*	*Estimated frequency*	*Observed frequency*	*Discrepancy*
No	No	No	96.3	94	2.3
No	No	Yes	62.7	65	−2.3
No	Yes	No	26.7	29	−2.3
No	Yes	Yes	81.3	79	2.3
Yes	No	No	29.7	32	−2.3
Yes	No	Yes	32.3	30	2.3
Yes	Yes	No	28.3	26	2.3
Yes	Yes	Yes	132.7	135	−2.3
			490.0	490	0.0

The fits of the model for both groups are good, and rather similar, with the high education group showing a somewhat better fit than the low.

The computed values of the model parameters for the two groups are given below, together with the previously given parameters for the two groups combined:

Time	*Latent probability*			*Proportion in latent class 1*		
	Total group	*Low education*	*High education*	*Total group*	*Low education*	*High education*
1	0.80	0.81	0.78	0.62	0.67	0.58
2	0.86	0.84	0.88	0.47	0.52	0.44
3	0.83	0.81	0.84	0.37	0.44	0.31

The differences between the two groups are rather small in terms of the latent probabilities, and they follow a rather similar pattern. However, even small differences here can make for surprisingly large differences on the manifest level, and are not without interest. At the initial interview the low education group showed a slightly higher latent probability, but at the second interview, while both groups showed increased latent probability, the high education group had increased much more to a higher level than the low. At time 3, the low education group dropped back to the latent probability (0.81) they had shown at time 1, but while the high education group dropped also, they retained a latent probability higher than that of time 1 (0.84 as against 0.78), and higher than that of the low education group at time 3.

It is in the latent changes (changes of latent proportions) that the greater differences emerged between the two groups. The initial difference in latent proportions in Class 1 between the low and high education groups was greater than the manifest difference, the usual situation (see next chapter), but the change between time 1 and time 2 in terms of percentage points was about the same for the two groups. The low education group dropped from 0.67 to 0.52, the high from 0.58 to 0.44. The movement continued in the same direction for both groups, with the low education group moving down eight more points at time 3 and the high education group moving down thirteen more points at time 3.

Keep in mind that a lower proportion means that more parents said that their children came to them with a problem. Without getting into other kinds of methodological problems, we find that the responses moved more in the direction desired by the program directors on the latent level than on the manifest, and that it continued to move more for the better educated parents than for the less educated.

Use of the Same Model with Data from a Consumer Anticipations Study

For years the important consumer organization Consumers Union of U.S. has conducted a variety of mail questionnaires with its subscribers. In 1958 the organization permitted certain economists to solicit panel data by mail from subscribers on a volunteer basis. One group involved was the National Bureau of Economic Research, whose work on this topic was largely directed and reported by F. Thomas Juster. In particular, the data sources and methods of collection are reported in some detail in Appendix C of Juster (1964), to which the reader is referred for this information. I am indebted to Juster for helpful suggestions about economic anticipations data and their interpretation.

However, most of Juster's work was limited to the study of two-wave panels. Albert G. Hart, then Chairman of the Economics Department and director of the Expectational Economics Center of Columbia University, carried the panel forward for two more waves. I am indepted to Hart for the use of data from this project.

For the present purpose, an example is chosen from the first three panel waves. The item referred to anticipations of the future of household income for the year following the date on which the question was answered. The three questionnaires were conducted in April 1958, October 1958, and April 1959 (see Hart, 1966, for details).

Specifically, the item used here asked the question "Do you expect that your family income during the next 12 months, compared to your income during the last few months, will be" followed by a series of code categories which referred to various degrees of possible increase, stability, or decrease. For the present analysis these categories have been dichotomized into the group of categories which refer to an anticipated increase in income of moderate or greater, and those which refer to an anticipated increase which would be slight, or to no increase or a decrease. This gave the dichotomization which was closest to a median split.

The manifest proportion positive was 0.56 at time 1, 0.52 at time 2, and 0.46 at time 3. This probably reflected a general decline in the economy which was widely expected at that time (and which in fact was occurring in many sectors of the economy).

With 12,104 cases giving responses on all three waves on this item, turnover was 3,081 cases between time 1 and time 2, 3,257 cases between time 2 and time 3, and 3,538 cases between time 1 and time 3. This pattern of turnover is somewhat different from that for many of the other examples, in that the turnover increased between waves 2 and 3, but it was generally similar in that the turnover between waves 1 and 3 was not a great deal larger than that between the single time intervals.

The wording of the question was changed slightly between waves, and this may have contributed to the slightly different pattern of turnover. The estimated and observed frequencies for the example are shown in Table 36.

The fit is not as close as we would perhaps like (for those interested, chi-square is 5.97), and it may be that this is not the proper model for this item. Although the respondents were of very

TABLE 36

Estimated and observed third order frequencies, responses to item "Do you expect your family income during the next twelve months, compared to your income during the last few months, will be moderately greater or more, or only slightly greater, the same, or less?", Consumers Union Panel, Columbia University

Time 1	*Time 2*	*Time 3*	*Estimated frequency*	*Observed frequency*	*Discrepancy*
High	High	High	3711.5	3738	−26.5
High	High	Low	1280.5	1254	26.5
High	Low	High	699.5	673	26.5
High	Low	Low	1060.5	1087	−26.5
Low	High	High	620.5	594	26.5
Low	High	Low	700.5	727	−26.5
Low	Low	High	576.5	603	−26.5
Low	Low	Low	3454.5	3428	26.5
			12,104.0	12,104	00.0

high caliber, in addition to being volunteers, there may have been some aspects of this study or of this item which were extraneous and which threw off the fit. It was noted above that the pattern of turnover was a bit different from that usually observed in these examples. Of course, with such a large sample the discrepancies would be expected to be numerically larger than for smaller samples. The fit seems to be close enough, in any event, to warrant giving the computed parameters of the model, which were as follows:

Time	*Latent probability*	*Proportion in latent class 1*	*Proportion in latent class 2*
1	0.87	0.58	0.42
2	0.87	0.53	0.47
3	0.87	0.45	0.55

Here the latent probabilities remained stable, while the latent proportions positive declined steadily. The stability of the latent probabilities indicates a lack of increasing structurization of the situation, and it also leads us to believe that a different model might fit better, one which we shall refer to subsequently.

Other Possible Models Involving Both Latent Change and Change in Latent Probabilities

One model which might seem promising, particularly in an example such as the previous one, is a model which holds the latent probabilities constant, as was done in earlier chapters, but which does not limit the probability of a positive response for members of Class 1 to equality with the probability of a negative response for members of Class 2. In other words, we would have

$$a_1 = a_2 = a_3 \neq b_1 = b_2 = b_3$$

while admitting latent change in one direction from time 1 to time 2 and in the same direction from time 2 to time 3.

The model structure would then look as follows:

Time	*Latent probability of positive response from*		*Proportion in latent class 1*	*Proportion in latent class 2*
	latent class 1	*latent class 2*		
1	a_1	$1-b_1$	$v_{111} + v_{112} + v_{122}$	v_{222}
2	a_1	$1-b_1$	$v_{111} + v_{112}$	$v_{112} + v_{122}$
3	a_1	$1-b_1$	v_{111}	$v_{112} + v_{122} + v_{222}$

Interestingly enough, this model does not come anywhere near fitting the previous example, not as close as the previous model. It should be noted that actual computations on these models are generally required to be carried out to six significant digits, because the smallest rounding error can cause a surprisingly large difference in the regenerated data. However, to spare the reader detail we have not generally reported more than two or three digits. Thus numbers which are given as identical are not identical if carried to a greater number of digits, but because of rounding.

The model above was also applied to all the other examples to which the previous model was applied, plus a great many more, and not a single good fit was found. The best fit obtained was to the item from Project ENABLE on knowledge of marriage conselling resources in the community. The model parameters obtained were a_1 = 0.83 and b_1 = 0.88 with the latent proportion positive declining from 0.71 at time 1 to 0.49 at time 2 to 0.45 at time 3. The principal differences between these parameters and those of the previous model for this item were that b_i, the probability that a person in Class 2 gives the negative response, is higher

TABLE 37

Estimates and observed third order frequencies, responses to item on knowledge of community resources for marriage counselling, ENABLE panel

Time 1 knowledge	*Time 2 knowledge*	*Time 3 knowledge*	*Estimated frequency*	*Observed frequency*	*Discrepancy*
No	No	No	256.6	241	15.6
No	No	Yes	94.4	100	− 5.6
No	Yes	No	75.4	81	− 5.6
No	Yes	Yes	181.6	186	− 4.4
Yes	No	No	57.4	63	− 5.6
Yes	No	Yes	45.6	50	− 4.4
Yes	Yes	No	41.6	46	− 4.4
Yes	Yes	Yes	232.4	218	14.4
			985.0	985	00.0

here than before (when by assumption it equalled a_i, and the a_i ranged over time from 0.82 to 0.84), and b_i is the higher latent proportion at all three times in the present model. The estimates and observed values of the third order frequencies are presented in Table 37.

The chief failure of this model on this example is that the model underestimates the amount of change which appears in the manifest data in any direction.

Numerous other models of similar sort may be conceived, and in fact have been constructed, solved, and applied to a considerable number of items from the studies discussed above. For example, the models above were all shown as applied to examples in which the manifest marginals moved in one direction. Now we know that the manifest and latent marginals do not have to move in the same direction, as shown in the example in a previous chapter of the Truman—Dewey campaign, where manifest opinion over one time period moved towards Dewey while latent marginals moved towards Truman. Hence we may think of applying either the models above or models permitting change in opposite directions over the two time intervals to data showing manifest change in opposite directions.

Applying the model with latent probabilities unrestricted over time (but with $a_i = b_i$), and with latent change in only one direction, to manifest data where the marginals moved in opposite directions, yielded nothing approaching a good data fit, for a large number of items.

Again, models with the same assumptions about latent probabilities but allowing latent change in opposite directions over the two time periods, even though they had seven unknowns — equal to the number of independent known quantities — produced nonsensical results, such as negative probabilities, when applied to the same items.

Another model required a_i to remain constant over time, b_i to remain constant over time but not necessarily equal a_i, and permitted latent change in opposite directions over the two time periods. Here again, meaningless results were obtained when the model was applied to a body of items, including many from the studies cited — such results as probabilities which were imaginery numbers, or greater than one, and so on.

It is perhaps as well that the failures to get fits are more numerous than the successes, since some conclusions can be drawn from the failures as well as the successes. Some of these conclusions will be stated in the following chapter, along with some other implications of these models, and some of the more general conclusions will be in the concluding chapter.

Analogues for Continuous Variables to the Models of this Chapter

There seems to be little profit in developing for continuous variables analogues to each specific model shown for discrete variables. We shall, however, briefly discuss the general implications of models with changing latent probabilities for correlation matrices.

Models Without Changing Latent Positions

The analogue to changing latent probability is changing reliability, where the term is broadly used to cover all random elements in observed scores. Let us use the notation r_{r_i} to refer to the reliability at time *i*. For the models where there is no change in latent position the true correlations between all pairs of times are equal to one. Since also the correlation of a true score with itself is equal to one, the correlation matrix of true scores has all its elements equal to one.

From (3.53) we have

$$r_{X_iX_j} = \sqrt{r_{r_i}}\sqrt{r_{r_j}}\, r_{T_iT_j} \tag{6.18}$$

This corresponds to the well-known formula for attenuation. Let us define a diagonal matrix as follows

$$R_r = \begin{pmatrix} \sqrt{r_{r_1}} & 0 & . & . & . & 0 \\ 0 & \sqrt{r_{r_2}} & . & . & . & . \\ . & . & . & . & . & . \\ . & . & . & . & . & . \\ . & . & . & . & . & . \\ 0 & . & . & . & . & \sqrt{r_{r_m}} \end{pmatrix} \tag{6.19}$$

where m is the number of time points or panel waves available. Let us call the matrix of true correlations, with all elements equal to one, R_T, and the matrix of observed correlations R_X. Then

$$R_X = R_r R_T R_r \tag{6.20}$$

by (6.18). Written out for $m = 3$, this is

$$R_X = \begin{pmatrix} r_{r_1} & \sqrt{r_{r_1}}\sqrt{r_{r_2}} & \sqrt{r_{r_1}}\sqrt{r_{r_3}} \\ \sqrt{r_{r_2}}\sqrt{r_{r_1}} & r_{r_2} & \sqrt{r_{r_2}}\sqrt{r_{r_3}} \\ \sqrt{r_{r_3}}\sqrt{r_{r_1}} & \sqrt{r_{r_3}}\sqrt{r_{r_2}} & r_{r_3} \end{pmatrix} \tag{6.21}$$

Solving this equation for the reliabilities is quite simple; for example, when $m = 3$, we have

$$r_{r_1} = \frac{r_{X_1X_2}\, r_{X_1X_3}}{r_{X_2X_3}}$$

or in general

$$r_{r_i} = \frac{r_{X_iX_j}\, r_{X_iX_k}}{r_{X_jX_k}} \tag{6.22}$$

In this model, the correlations in any row or column are proportional to the square root of the reliability for the time corresponding to the row or column. If the reliabilities tend to increase through time, all the correlations in successive rows and columns will tend to increase. Let us use a schematic representation of the observed correlation matrix, with arrows indicating increasing correlations. Then if the reliabilities increase through time, the matrix generated by the model of this section would look like this:

Most of the available data, however, look like this:

Hence this model will not account for much of the available data.

Models With Changing Reliabilities and Changing Latent Positions

Here the reliability may change and the true score may change. The equations (6.18), (6.19), and (6.20) still hold, although now the matrix R_T of true correlations does not consist entirely of elements equal to one. It now has elements equal to one in the principal diagonal, and true correlations, less than one in general, as all other elements. Written out for $m = 3$, the relationship is

$$R_X = \begin{pmatrix} r_{r_1} & r_{T_1T_2}\sqrt{r_{r_1}}\sqrt{r_{r_2}} & r_{T_1T_3}\sqrt{r_{r_1}}\sqrt{r_{r_3}} \\ r_{T_2T_1}\sqrt{r_{r_2}}\sqrt{r_{r_1}} & r_{r_2} & r_{T_2T_3}\sqrt{r_{r_2}}\sqrt{r_{r_3}} \\ r_{T_3T_1}\sqrt{r_{r_3}}\sqrt{r_{r_1}} & r_{T_3T_2}\sqrt{r_{r_3}}\sqrt{r_{r_2}} & r_{r_3} \end{pmatrix} \tag{6.23}$$

These equations cannot be solved for the reliabilities without some further assumption. We might, for example, assume that the matrix of true correlations has unit rank, i.e., that true scores at time t are independent of true scores at time $t - 2$ when true scores at time $t - 1$ are controlled. This was the assumption made in one model of Chapter 3. With that assumption we have relationships in the true matrix of the type

$$r_{T_iT_{i+2}} = r_{T_iT_{i+1}} \cdot r_{T_{i+1}T_{i+2}}$$

We thus get, from the observed correlations, relationships of the type

$$\begin{vmatrix} r_{X_iX_{i+1}} & r_{X_iX_{i+2}} \\ r_{r_{i+1}} & r_{X_{i+1}X_{i+2}} \end{vmatrix} = 0 \qquad (6.24)$$

For any given i, with m large, there will be a number of such determinantal solutions, except for $i = 1$ and $i = m$, for which cases there will not be solutions. The situation is exactly like that of the latent distance models of latent structure analysis, where the cross-products [ii] of items with themselves can be estimated for all but the end items.

We might note at this point that all of the models of this paper, since they deal with time series or panel data, have to reckon with the fact that the data have a natural order. The data collected at a particular time come before the data collected at another particular time and after the data collected at some other time, and the order cannot be changed. If we look to scaling theory and factor analysis for mathematical similarities, we must look to scales in which the items are ordered, such as latent distance scales and Guttman scales or to theories of order factors, such as Guttman's radex theory (see Guttman, 1954).

Since each reliability, if m is large, has a number of solutions of the type of (6.24), we need to take an average of some sort in order to obtain an estimate. Here exactly the same computing equations as those of latent distance models can be used, and the details will not be given.

Example of this Model

We shall use the same variable from the class panel which was employed as an example in Chapter 3, which we called Goal-Aspiration, or, more generally, Level of Aspiration, except that now we shall use all ten waves of the panel rather than the first four. We assume that the matrix has unit rank and permit the reliabilities to change, computing a separate value for the reliability at each time except the first and last. The data are given in Table 38.

The series really splits into two parts of five waves each, because the first semester of the course ended after wave 5, and it can be seen from the correlations that the relation of wave 6 (immediately prior to which the students received grades for the first semester) to prior waves is idiosyncratic. With this exception and with a

TABLE 38

Goal aspiration of sociology students, matrix of product—moment correlations

Time	1	2	3	4	5	6	7	8	9	10
1	/	0.71	0.67	0.62	0.59	0.49	0.58	0.48	0.57	0.51
2		/	0.85	0.79	0.77	0.58	0.68	0.59	0.67	0.64
3			/	0.90	0.82	0.63	0.71	0.64	0.68	0.58
4				/	0.87	0.69	0.78	0.66	0.71	0.63
5					/	0.72	0.76	0.68	0.72	0.64
6						/	0.90	0.78	0.82	0.75
7							/	0.86	0.89	0.82
8								/	0.91	0.84
9									/	0.91
10										/

N = 71 (except time 1, 2, and 3, when N = 66)

few anomalies not surprising with so few cases, the data display a fairly regular "roof-top" pattern.

When the average estimates based on equations of the type of (6.24) are computed, the following figures result:

TABLE 39

Estimates of reliability, goal aspiration, class panel

Time	1	2	3	4	5	6	7	8	9	10
r_{r_1}	—	0.88	0.88	0.90	0.86	0.78	0.96	0.85	1.00	—

We do not find any pattern here, probably in part because the matrix does not really have unit rank across the semesters.

CHAPTER 7

Some Further Implications and Problems of Models

There is of course an enormous range of implication for each class of models considered here, and this chapter will consider some of the more important methodological implications of various classes of models, important also for substantive interpretations.

First, let us look at the effects on marginals of dealing with manifest items which are only probabilistic in their relation to latent variables. In psychological test theory, traditionally, the assumption was that the error of measurement had a mean of zero. Thus for a sample of some size the true and observed means were equal.

However, in dealing with attribute data, such is not in general the case. The effect on marginals depends on the latent marginal as well as the latent probabilities. If p_1 is the manifest positive marginal, a_1 is the probability of positive response at time 1 for latent class 1, and b_1 is the probability of positive response at time 1 for latent class 2 (changing the notation slightly and assuming only two latent classes), then at time 1 we have the following relationship:

$$\begin{aligned} p_1 &= a_1 v_1 + b_1(1-v_1) \\ &= b_1 + (a_1-b_1)v_1 \end{aligned}$$

TABLE 40

Values of p_1 when true proportion in latent class 1 and latent probabilities are given (for dichotomous items)

v_1	$a_1 = 0.80$ $b_1 = 0.20$	$a_1 = 0.90$ $b_1 = 0.10$	$a_1 = 0.95$ $b_1 = 0.05$	$a_1 = 0.90$ $b_1 = 0.20$	$a_1 = 0.95$ $b_1 = 0.10$	$a_1 = 0.95$ $b_1 = 0.15$	$a_1 = 0.9$ $b_1 = 0.1$
1.00	0.80	0.90	0.95	0.90	0.95	0.95	0.98
0.90	0.74	0.82	0.86	0.83	0.865	0.87	0.90
0.80	0.68	0.74	0.77	0.76	0.78	0.79	0.82
0.70	0.62	0.66	0.68	0.69	0.695	0.71	0.74
0.60	0.56	0.58	0.59	0.62	0.61	0.63	0.66
0.50	0.50	0.50	0.50	0.55	0.525	0.55	0.58
0.40	0.44	0.42	0.41	0.48	0.44	0.47	0.50
0.30	0.38	0.34	0.32	0.41	0.355	0.39	0.42
0.20	0.32	0.26	0.23	0.34	0.27	0.31	0.34
0.10	0.26	0.18	0.14	0.27	0.185	0.23	0.26
0.00	0.20	0.10	0.05	0.20	0.10	0.15	0.18

Table 40 shows some examples of values of p_1 over a range of values of v_1 (the latent positive marginals) for selected pairs of values of a_1 and b_1. The horizontal lines in the columns show the interval where p_1 will approximately equal v_1 for the given a_1 and b_1. Above the horizontal line (or lines) p_1 will be an underestimate of the true value v_1; below the line (or lines) p_1 will be an overestimate of v_1.

In the first three columns, where $a_1 = (1 - b_1)$, the differences are symmetrical around the value $v_1 = 0.50$. But when a_1 is greater than $(1 - b_1)$, p_1 overestimates v_1 for certain values of v_1 greater than 0.50, and the difference depends upon the values of a_1 and $(1 - b_1)$ and the difference between them.

Now we have seen that in much of the data the latent probabilities tend to increase through time. It is reasonable to suppose, for example, that in the political data they increase until the election and then decrease until the next political campaign starts. If we consider two groups of respondents, one with a true proportion positive of $v_1 = 0.80$ and the other with $v_1 = 0.20$, the following table shows what would happen to the manifest positive marginals for each group as the latent probabilities rise and fall (assuming that the latent probabilities of the two latent classes are complementary).

	Before campaign starts	*Begin-ning*	*Climax*	*Elec-tion*	*After-math*	*Dol-drums*
a_1	0.70	0.80	0.90	0.95	0.80	0.70
b_1	0.30	0.20	0.10	0.05	0.20	0.30
Group 1 ($v_1 = 0.80$)	0.62	0.68	0.74	0.77	0.68	0.62
Group 2 ($v_1 = 0.20$)	0.38	0.32	0.26	0.23	0.32	0.38

The authors of the Elmira study (Berelson *et al.*, 1954) suggest that some such pulsation occurs between groups, in fact. A polarization occurs during the campaign, and recedes between elections. The present models can account for the phenomenon, but it should be kept in mind that there is no necessary connection between these models and the particular theorizing about causes of the phenomenon in which the authors of the Elmira volume engage.

If the latent probabilities are not complementary, and further if they change at a different rate or in a different way, we may expect even more uneven departures of the manifest marginals from the latent marginals. Such relationships undoubtedly account for many of the strange departures which political polls often take from the election results, where in other situations they are very accurate. The concluding chapter will have some further comments on this subject, but we can see at a glance that a small extreme hard-core group is likely to hold up better at the actual election than a broad thin group with low latent probability, other factors being equal.

Second, let us consider the effect of a change in latent probabilities on the relation between two items. Let us use for this purpose the notation p_{ij} for the proportion of the sample reported in category i on item a and category j on item b. v_{ij} will be the proportion in the corresponding latent categories. a_{rs} will be the probability that a respondent in the latent class C_r on item a will be reported in category s on item a. b_{rs} will be the probability that a respondent in latent class C_r on item b will be reported in category s on item b. We have the equation

$$p_{11} = a_{11}b_{11}v_{11} + a_{11}b_{21}v_{12} + a_{21}b_{11}v_{21} + a_{21}b_{21}v_{22}$$

Suppose we define a perfect latent association between the two items as a distribution in which v_{12} and v_{21} are both zero, so that

$$v_{11} + v_{22} = 1$$

Then we have

$$p_{11} = a_{11}b_{11}v_{11} + a_{21}b_{21}v_{22}$$
$$= a_{11}b_{11}v_{11} + (1-a_{22})(1-b_{22})v_{22}$$

Now if $v_{11} = 1$,

$$p_{11} = a_{11}b_{11}$$

Since in general $a_{11} > 1 - a_{22}$ and $b_{11} > 1 - b_{22}$, this equation represents the maximum p_{11} can reach, i.e., p_{11} will not be greater than the product of the latent probabilities of category 1 for both items at both times.

If $v_{11} = 0$,

$$p_{11} = (1-a_{22})(1-b_{22})$$

Since $1 - a_{22} < a_{11}$ and $1 - b_{22} < b_{11}$, this represents the minimum that p_{11} can be when the latent association is perfect. In other words, when the latent association is perfect,

$$(1-a_{22})(1-b_{22}) < p_{11} < a_{11}b_{11}$$

Similarly,

$$(1-a_{11})(1-b_{11}) < p_{22} < a_{22}b_{22}$$

More generally, define the manifest and latent joint occurrence matrices as follows:

$$P_{ab} = \begin{pmatrix} p_{11} & p_{12} \\ p_{21} & p_{22} \end{pmatrix} \qquad V_{ab} = \begin{pmatrix} v_{11} & v_{12} \\ v_{21} & v_{22} \end{pmatrix}$$

Also let

$$A = \begin{pmatrix} a_{11} & a_{12} \\ a_{21} & a_{22} \end{pmatrix} \qquad B = \begin{pmatrix} b_{11} & b_{12} \\ b_{21} & b_{22} \end{pmatrix}$$

Then

$$P_{ab} = A'V_{ab}B \tag{7.1}$$

If A and B have been determined by the methods of this paper or any other method, the joint latent distribution can be found by the equation

$$V_{ab} = A'^{-1}P_{ab}B^{-1} \tag{7.2}$$

If one takes, say the determinant of a 2×2 matrix as a measure of association between items, (7.1) can be put into the form

$$|P_{ab}| = |A'| \, |V_{ab}| \, |B| \tag{7.3}$$

Since $|A'|$ and $|B|$ must each be less than 1 if the latent probabilities are less than 1, the effect of latent probabilities on the latent association is obvious from this equation. If it is desired to compute the true association directly, we have

$$|V_{ab}| = \frac{|P_{ab}|}{|A'| \, |B|} \tag{7.4}$$

Now suppose that through time the association between the two sets of latent classes does not change at all, but the latent probabilities of items a or b or both increase. This means that the diagonal elements of the matrices A and B increase, so that the determinants of these matrices become larger (they are always positive). Then by (7.4) the determinant of p_{ab}, which is a measure of association of the two sets of manifest responses, will increase.

In other words, in a set of items all of whose latent probabilities are increasing, the associations among all the items will increase, even though the relationships between pairs of sets of latent classes does not change. Similarly, if the latent probabilities decrease, the manifest associations among the items will decrease.

Now it is a well-confirmed fact in political panel studies that the correlations among most of the related items which can change through time increase as the campaign wears on. This phenomenon has not been satisfactorily explained, although a number of alternative hypotheses are available from social psychology. More will be said about this in the concluding chapter, but here it may be noted that all that is required for an explanation is any process which brings about an increasing probability between the latent and manifest positions on some of the items, without the necessity for referring in any way to the substantive relationships among the items. Questions such as cross-pressures, dissonance, causation, and so on, need not be invoked. The matter may be much simpler than has been discussed.

Skewness and Homogeneity

One of the beauties of the normal distribution is symmetry. In panels of the kind we are considering, and with variables such as are being used constantly in the fields of application with which we deal here, we are likely to have items which are far from symmetrical on either the latent or the manifest level.

Beyond that, we are unlikely to be using items in which latent classes are genuinely homogeneous with respect to their latent probabilities. The matter is further complicated by the likelihood of a change in the shape of the latent distribution over time, and even a change in the space. Often, we even find panels in which the space of the manifest responses has been changed over time!

There is a genuine Hobson's choice between parsimony in a model, sufficient to leave at least a little freedom for even intuitive notions of a decent fit to play a part, and the almost infinite complexity of the situations with which we are dealing.

For example, the writer has always found graphic thermometers (except where a small number of qualitative attributes were clearly what was needed) to be the form of manifest item which yielded the most information for the least researcher—respondent effort. These manifest scales take various forms, and the distributions which are obtained range from approximately normal, highly skewed unimodal, rectangular, to U-shaped. They are tapping latent variables which are at least as complicated. So, for that matter, are a great many deceptively simple dichotomous items and other seemingly simpler manifest forms.

Further to complicate matters, when these items are looked at over time, say in the form of a three-wave panel, the change distribution is very often found to be quite different (in fact it is usually quite different) from the distribution at any one time. In particular, the change distribution is very frequently U-shaped — that is, there is a relatively inert mass at each end of the distribution and a smaller number of changes in between. On the other hand, there are instances of distributions which are U-shaped at each separate time but whose changes show approximately a normal distribution.

Especially in field studies, non-experimental data, we are always dealing with approximations. Often there is lack of systematicity at crucial points. Often the assumptions are known to be at some degree of variance with the facts. Science, however, frequently develops by what often is called "the method of successive approximation."

One important methodological point in respect to the present models arises from grouping responses into a limited number of categories in which latent probabilities are assumed to be equal for all occupants of the category at a given time. Typically, for parsimony, we have been assuming latent dichotomies, and have been using manifest dichotomies, either taken from direct responses or created by grouping. Generally, we have found the best fits to be models which assume that the latent probability for a person in the positive latent class to fall in the positive manifest class is equal to the probability for a person in the negative latent class (or class 2) to fall in the negative manifest class.

The point to be made here is that the approximation based on the assumption of equal latent probabilities depends among other things very heavily on the skewness of the item. In other work on product images, the writer has shown that the amount of turnover of these images depends (inversely) on item skewness (which in that case is a measure of saliency), created skewness (which depends on a manipulation of the grouping of manifest responses), and on ownership of the product and the type of image (Wiggins, 1958; see also Lazarsfeld, 1954). Despite the fact that we are here referring to latent probabilities rather than turnover, we know that the turnover largely reflects the latent probability, with low latent probability resulting in high turnover, apart from any substantive factors.

Numerous examples have been computed which demonstrate the present point, but since highly skewed cuts do not usually yield good results they have not often been reported. One may be selected from the ENABLE Panel, an item for which an excellent fit was found for the median split of a graphic thermometer. This item, reported in Chapter 6, was "I yell at my children more than I should," and the positive response was taken to be the high end of the scale. The dichotomization at the median yielded a manifest proportion positive of 0.53 at time 1, 0.47 at time 2, and 0.46 at time 3.

We have now split the item into a dichotomy in which as near as possible to one-third of the responses are, on the average over time, on the high side. This yielded manifest proportions "high" of 0.39 at time 1, 0.31 at time 2, and 0.28 at time 3. The amount of turnover from time 1 to time 2 is 301 cases, from time 2 to time 3 it is 249 cases, and from time 1 to time 3 it is 300 cases. This is smaller than for the median dichotomization, and corresponds with the findings cited above (Wiggins, 1958).

The model which was used required the latent probabilities of the two classes to be complementary at a given time, but they

TABLE 41

Estimated and observed third order frequencies, responses to item "I yell at my children more than I should," with dichotomous cut nearest one-third "high," ENABLE panel

Time 1	*Time 2*	*Time 3*	*Estimated frequency*	*Observed frequency*	*Discrepancy*
High	High	High	137.9	130	7.9
High	High	Low	56.1	64	–7.9
High	Low	High	45.1	53	–7.9
High	Low	Low	145.9	138	7.9
Low	High	High	30.1	38	–7.9
Low	High	Low	79.9	72	7.9
Low	Low	High	67.9	60	7.9
Low	Low	Low	427.1	435	–7.9
			990.0	990	0.0

could move freely over time, and the latent change in one direction was allowed for the first interval and in the same direction for the second interval. The results of applying the same model to the same cases with the different dichotomous cut, giving estimated and observed third order frequencies and discrepancies, as before, are shown in Table 41.

Again, let us look at the results of a dichotomization of the scale in which the nearest to two-thirds of the responses, on the

TABLE 42

Estimated and observed third order frequencies, responses to item "I yell at my children more than I should," with dichotomous cut nearest two-thirds "high," ENABLE panel

Time 1	*Time 2*	*Time 3*	*Estimated frequency*	*Observed frequency*	*Discrepancy*
High	High	High	479.4	490	–10.6
High	High	Low	86.6	76	10.6
High	Low	High	75.6	65	10.6
High	Low	Low	59.4	70	–10.6
Low	High	High	93.6	83	10.6
Low	High	Low	33.4	44	–10.6
Low	Low	High	32.4	43	–10.6
Low	Low	Low	129.6	119	10.6
			990.0	990	00.0

average, are placed on the "high" side. This yielded manifest proportions "high" of 0.71 at time 1, 0.70 at time 2, and 0.69 at time 3. The turnover was 262 cases between time 1 and time 2, 228 between time 2 and time 3, and 272 between time 1 and time 3. Again, the turnover is lower than for the median dichotomization, and it also turns out to be lower than for the previous cut with one-third "high".

The observed and estimated third order frequencies and discrepancies are shown in Table 42.

The latent model parameters are as follows:

Time	*Latent probability*			*Proportion in positive latent class*		
	Median cut	*One-third high cut*	*Two-thirds high cut*	*Median cut*	*One-third high cut*	*Two-thirds high cut*
1	0.82	0.88	0.84	0.55	0.35	0.81
2	0.84	0.85	0.88	0.46	0.22	0.77
3	0.82	0.87	0.87	0.44	0.21	0.76

The average discrepancies of these two models, about 7.9 for the first and 10.6 for the second, are much greater than the 1.3 found for the median dichotomization. The assumption of equal latent probabilities for all members of each class is a much better approximation, other things being equal, when we are cutting into a distribution in such a way that the position of members of each category average out about the same in terms of the variable in relation to the cutting point. A median dichotomization takes care of a great many unknown variables. The same strictures apply, mutatis mutandis, to items whose response categories are formulated as dichotomous attributes, or in any other form, for that matter.

As a further empirical confirmation of the point, we have ranked the seven sets of data used in the examples from the medical project, the Sandusky, and the Elmira voting studies, in terms of the closeness at time 2 in each example to a fifty-fifty split in the manifest data. We then ranked the best-fitting model used above for these examples for each example, on the basis of discrepancies, admittedly somewhat impressionistic in the absence of statistical tests of goodness of fit, but nevertheless a rank order with which most readers would agree.

The Spearman rank correlation coefficient was 0.79. In other words, the greater the skewness in the responses, the worse the fit.

This covers a range of models, and is certainly not conclusive. However, the point remains.

This leads to the general methodological suggestion that when a manifest dichotomy is to be used and it is possible to group the categories in various ways, a clearer picture is likely to emerge when the responses are grouped so as to be as near the median as possible than from other groupings. Of course, it is not necessary to use only two manifest categories, and indeed explicit mathematical solutions for many models using any number of manifest response categories are given in Appendix B. The number used will depend on substantive, theoretical, and cost considerations (including sample size), some of which are discussed in the first and last chapters of this monograph.

CHAPTER 8

Conclusion

Crass Casualty obstructs the sun and rain
Thomas Hardy

No one knows, and it probably could not be estimated, how many hundreds of millions of panel items of social and economic behavior are collected from various human samples and populations every year in the United States alone — there may be billions — although many of them are not analyzed in terms of changes. The bulk of them are, however, analyzed as single items when they are analyzed in terms of changes, or they are put together in indexes, test scores, scales, and so on, which are in turn analyzed as single variables.

This is not to say that some form of analysis of the simultaneous movement of a number of variables, as in certain simulations and mathematical models, would not yield more information. However, there is a great deal to be learned from the analysis of the movement of single variables, which has barely been begun.

Indeed, if we cannot clarify the movement of a single variable to some extent, it is doubtful whether the more complicated movements involving a whole group or system of variables can be clarified. This may not always be true, but it is likely to be the case where we do not have tight theoretical formulations and specifications of what variables are relevant. That is the general situation in the areas covered here. For example, to my knowledge no population experts forecast early on the sharp decline in birth rates and expected birth rates now occurring in the United States, despite immense work by very good minds in this area. The discrepancies in forecasts among various econometric models over the past decade are well known.

Admitting that a problem is large and important does not necessarily make it easy or even possible to solve, at a given time. Conversely, what appears to be nit-picking to many — as witness the law of falling bodies — may turn out to have considerable significance.

In this essay I have addressed myself to four aspects of certain models dealing with the movement of a single variable taken from a sample or population of human subjects at several points in time. (I have briefly touched on a few implications for the relationship of two or more variables.) These four aspects are mathematical, methodological, empirical, and theoretical, the last least of all in what has gone before. In this final chapter I propose briefly to summarize what has been done in the first three areas and then to amplify somewhat on the last. Of course, the four areas overlap in many ways.

Mathematical Models

There are evidently innumerable models for univariate panel data which might be developed. Most of the non-experimental data in the social sciences comes in the form of observations discrete in time, and the great majority show amounts of total change which in the absence of important exogenous disturbances (and sometimes even with them) are not much greater for a longer time period than for a shorter.

These facts have led here to the proposal for a number of mathematical models based on the latent—manifest distinction, using both discrete and continuous spaces but only discrete times. The latent—manifest distinction used here is very similar though not identical to that of Lazarsfeld. As used here it includes the following basic propositions.

(1) For any given variable, there are a number of possible latent positions a respondent or subject might occupy.

(2) For any given latent position (or interval, for a continuous latent space), there is a probability that a person holding that position will show, in recorded data, a given manifest position, as measured by whatever instruments are used.

(3) In general, there is no necessary relationship between the

type of latent space and the type of manifest space. For example, a dichotomous manifest item may tap a latent continuum and vice versa. This in itself distinguishes the latent—manifest relationship from the classical approaches to reliability of a test.

(4) For any given empirical set of data, there will be at least two basic kinds of parameters. One is the set of probabilities that a person holding a given latent position will occupy any particular manifest position (or fall within a given interval); the other is the proportion of respondents or subjects who occupy a given latent position or interval.

(5) Either or both of these two kinds of parameters may remain unchanged over time or may change.

(6) If either or both kinds of parameters change, they may change systematically or unsystematically in terms of some model of process or change. The simplest system, of course, is no change; but any kind of process can occur on the latent level. The discrete Markov chain has been considered above, but models of other sorts could just as well have been considered.

(7) The combination of no change, systematic change of some sort, or unsystematic change on the part of the two kinds of parameters has led to a classification into nine types of models, as shown in Chapter 1 above. The use of both discrete and continuous space provides a further classification.

As it happens, some of the possible models have been developed in detail; others scarcely at all. Many special models within some of these classes have also been applied to available bodies of data. Some models have been set forth and not explicitly solved mathematically. Generally, if there are more unknowns than independent items of data, models cannot be solved. The general problem of identification has not been pursued. In some instances we have over-identification; or alternatively we can use smaller amounts of data to get a solution and then test the goodness of fit. No attempt has been made to develop a statistical approach; where there are degrees of freedom an intuitive approach has been used. This, of course, is a highly controversial field, especially because we are rarely dealing with good probability samples from a well-defined universe, but often with biased samples, finite populations, and so on.

Enough algorithms have been presented to show that in principle most of these models can be solved mathematically if sufficient data are available in a particular problem. Further, they can be extended and solved in various directions, always useful, however, only if sufficient data are available.

Methodological Considerations

Once the basic probabilistic relationship involved in the latent—manifest distinction is accepted — and it is here asserted that it does in fact hold to a varying degree in the fields of applicability, enormous in practice, which are discussed in this essay — then the methodological implications have an exceedingly wide range, overlapping many other kinds of methodological problems. These have been somewhat skimped in the text in order to get the main ideas out.

As previously noted, there are three main uses of panel data: (1) to cumulate data on individuals, along with purifying or cutting down elements on invalidity, unreliability, and so on — the monumental work of, e.g., Ferber's group (1959) is one of the best examples of this, along with that of the Census Bureau, the University of Michigan group, and others; (2) to obtain aggregate data in fields such as agricultural forecasts, investment plans of industries, and the like, as in the work of the Katona group at Michigan, the National Opinion Research Center, University of Chicago, the National Bureau of Economic Research, and so on; and (3) to study change on the level of the individuals, as in studies of buying behavior, learning, etc., or of differentiated groups of individuals, as in Newcombe's, Lazarsfeld's, Blumen's *et al.* studies, Rosenberg's occupational plans studies, and much subsequent work.

In terms of sheer volume, the last has probably been the least. The first approach, in effect, tries to take the latent position to the level of a reasonably reliable manifest indicator. Its value cannot be questioned, nor can that of the second approach, which may be the oldest.

Nevertheless, the third type of applicability has great significance in its own right, and its own variety of unique problems. A tremendous amount of potentially valuable findings may have been ignored by trying to eliminate what for the purpose at hand was garbage (or merely irrelevant to the problem), even though a detailed examination of the changes may have yielded other kinds of information of genuine value. Certainly political pollsters for candidates, and even for media, have not failed to inspect such changes. The problem there has generally been a shortage of funds and possibly a lack of sophistication in the analysis in some instances.

The present study really began in 1948 with the methodological problem which had been posed by McNemar (1946), that of separating the 'true" changes of a variable from the "error" changes caused by unreliability. It quickly became apparent that the element of randomness was not merely a matter of error of measurement, but was inherent in the behavior being studied. Some other investigators have subsequently addressed themselves to this precise problem, notably Coleman (1964).

However, most investigators have still not realized the methodological significance of the fact of randomness inherent in the behavior, and have made little attempt to deal with it. The fact is that when all the elements normally thought of as measurement errors are removed by great care in questionnaire and interview construction, verification, card cleaning, and so on, we still usually end up with a random factor which yields a turnover or total amount of individual change considerably in excess of the amount of change of a non-random character, especially in non-experimental attitude—opinion data but often in quasi-experimental studies even with what would appear prima facie to be "harder" data.

The fact that we ordinarily cannot assign a particular manifest change in a particular respondent to one or the other type of change by no means vitiates the usefulness of making the separation on a probabilistic basis. Further, the fact that we may call a change "random" and ascribe it to a latent probability, using various assumptions from one situation to another about temporal partial associations on the latent level being equal to zero, does not reduce the importance of knowing what these basic parameters are and how they do or do not vary in certain situations, defined substantively on the one hand and theoretically on the other.

It must also be constantly kept in mind, as noted before, that the concept of latent probability in no way implies indeterminacy in the philosophical sense. Consider flipping a "balanced" coin for heads or tails. If this is done a number of times, we get something approximating the binomial distribution, and if a number of long series are produced we begin to converge on the normal distribution. All this in no way implies that the outcome of a single toss is uncaused or indeterminate. In fact, if we had all the necessary minutiae about the finger motions involved in the toss, the atmospheric conditions, the surface on which the coin would fall, and so on, we could in principle predict the outcome. We do not do this because the level and complexity of explanation required would be both beneath our interest and beyond our present capabilities of measurement. Instead, we simplify the problem and make useful inferences by assuming that all these minute elements add up to a random variable.

We may quote Tolman (1932) on this point:

> And, even should it finally turn out on a basis of further experiments, that there is for the behavior of organisms, just as for the behavior of electrons, some principle of ultimate indeterminateness (i.e., a kind of Heisenberg's uncertainty principle), this need not lead us to assume or suppose any metaphysically 'other' as 'butting in' to the course of organic nature. The finding of such an uncertainty principle would, to be sure, mean important and exciting things. It would mean that we must talk in terms of probabilities, of statistical averages, rather than in terms of unique individual cases. It would not mean, or at any rate would not need to mean, however, any metaphysical bifurcation or dualism — any breakdown in the possibility of final deterministic descriptions per se.

Thus the basic problem is not philosophical or terminological but practical. The fact is that in most panel studies of change the host of minutiae which in the aggregate make up the level of latent probability are not worth the immense cost which would be required for their detailed investigation, after obvious measurement errors have been removed. At the same time, there is going to continue to be a vast amount of data collection of single panel variables. And here, it is of great importance to have some knowledge of the contribution of the random element at any given time, since otherwise the interpretation of findings is almost certain to be erroneous (although not always to a degree which yields qualitatively erroneous conclusions).

If the changes in latent positions are over-riding in magnitude, the break-down may not be very important; but if, as is usually the case, the apparent turnover is mainly the product of the latent probabilities, and particularly if they are changing, as they often are, then any attempt to ignore this element of randomness and make a direct interpretation of the manifest data in a non-experimental study is likely to be grossly wrong. The difficulty here is finding the right model for the situation being studied, which as we have seen is no simple matter.

We have noted, among other methodological considerations, the fact that changes in latent probabilities alone can produce such commonly found occurrences as polarization, increasing or decreasing associations among variables over time, and apparent differences in brand loyalty or media exposure. The last, in fact, does not require changes in the latent probabilities. By separating the two sets of parameters we are able to make a meaningful analysis of the actual developments on the latent level, where they have different implications for action. We saw that in the 1948 political campaign the manifest data showed a slight trend to Dewey in Elmira while on the latent level the trend was to Truman, whom everyone predicted on the basis of manifest data would lose the election.

We have also had occasion to discuss the skewness of a variable as an important methodological consideration. Since the assumption of homogeneity of discrete groups with respect to latent probability is only an approximation, it is usually met better if it is possible on the manifest level to divide the sample into approximately equal groups, thus getting a better averaging effect. With many items this is not possible, and it will generally be found that the amount of turnover is smaller on the manifest level for skewed variables, and this in turn is often reflected on the latent level. More will be said about this below.

In general, the largest number of errors and the greatest difficulties of administration occur in the field work of non-experimental studies rather than in-house. At the same time, the cost of field work continually increases, while the cost of computer work continually declines. Thus although in principle field experiments are preferable, in practice non-experimental field work is going to continue to grow, and it is desirable to conduct as much research as possible in a great many areas, where reasonable models can be used through in-house model analysis. Thus many of the difficulties which arise in the kind of analysis proposed here will prove in the end to be smaller and cheaper to deal with than more elaborate field designs, wherever the latter are not absolutely necessary. Various forms of quasi-experimental designs will also be used more and more extensively (see, e.g. Campbell and Stanley, 1963, and Wiggins, 1954c).

We have scarcely touched upon many important matters, methodological and substantive, some of which will be referred to again in the concluding section. Among these are the effects of models of the type shown on the relationship among different variables, and the treatment of a system of variables moving over time, which is straightforward but would require more additional pages than we have already used. Also, such matters as floor—ceiling effects, regression on the mean, the fact that one manifest item may tap many latent variables (just as many manifest items may tap one latent variable), re-interview effects, memory, missing data, validity, the whole question of sampling variation, the use of finite populations — a host of methodological considerations have been mentioned barely or not at all. Questions of causation, of system, of endogenous effects in a system of variables (which would likely act on the changes of latent position) and of exogenous effects (which would act on both kinds of changes) have been left unexplored, to a great extent.

Empirical and Substantive Considerations

As noted, billions of panel items are collected each year in the United States alone. The areas where they are collected include, among them, the following:

(1) Social Security data — occupation, income, and the like.
(2) Income tax data.
(3) Dossiers of various sorts collected at all levels of government on millions of government employees, etc.
(4) As part of (3), records of millions of members of military service.
(5) Political opinion studies of monumental proportion.
(6) Economic studies of anticipations of individuals and firms.
(7) Media studies of thousands and perhaps millions of persons.
(8) Population planning studies.
(9) Medical studies, both private and public.
(10) Educational studies involving millions of students, their records, etc.
(11) Company records of performance, etc., of millions of workers.
(12) Union records of members with regard to payment of dues, etc.
(13) Dossiers on millions of convicts, suspects, people with alleged and sometimes actual subversive histories, etc.
(14) Monumental demographic studies.
(15) Evaluations of vast government programs in many other fields.

Most of these panel items are not used for the systematic study of change, in the sense dealt with in this monograph. The data are sometimes classified, proprietary, or confidential for other reasons. Nevertheless, there are billions of items on record, some hard, some soft, and often they could and perhaps should be dealt with in terms of change.

In this study we have dealt with bodies of data generally available and open for public inspection, particularly those at hand to the writer. No claim is made that they are intrinsically more important than others which might have been obtained. The changes in a farmer's prediction of the local soy-bean production are not necessarily less important than the reading level of a fourth-grade student in Ocean Hill—Brownsville, nor more so.

We have used data from media studies, consumer research, political studies, attitude—opinion surveys, educational research, poverty studies, marketing studies, and the like. They were obtained variously by government agencies, universities, private firms, consumer organizations, other eleemosynary organizations, and became available in a variety of ways. The amount of change which shows up in the manifest data varies greatly according to subject-matter; data collection procedures, including field procedures, type of instrument, and data processing and reduction procedures; the type of latent process or movement occurring; and a host of other factors. If there is latent movement the time interval may be important, but in most examples studied that kind of change tends to be swamped by movement occurring as a result of the latent—manifest probabilistic relationship at each individual point in time. Typically, we get in discrete dichotomous manifest measurement data from five to thirty per cent turnover from one time to the next.

Without the attempt to disentangle the turnover resulting from change on the latent level from that occurring as a result of the latent—manifest probabilistic relationship, many substantive errors of interpretation are made, and many seeming anomalies and paradoxes occur on the manifest level. Among others may be cited a study of a "United Nations Information Campaign", done over a week's time in Cincinnati in 1947, re-analyzed at the Bureau of Applied Social Research at Columbia University a year or two thereafter. This study was based on a barrage of media advertisements which pretty well saturated the media in Cincinnati for one week, attempting to provide information to people at large about the then new United Nations. A before—after panel was used to study the effects, and to the surprise of many, a large number of respondents showed a definite loss of information between the first and second panel waves. I later further analyzed this problem somewhat (although not in terms of the present models; see Levenson and Wiggins, 1954); but it is clear that the present models would (had there been enough panel waves for the analysis) have been more appropriate. The changes which occur as a result of the manifest—latent relationship (except when the latent probabilities are changing) are as indifferent to time order as regression on the mean, a fact of which Campbell has made so much good use.

Again, in studying economic anticipations, a question is often put on large-ticket items which asks "When do you intend to purchase a ....?" In a re-analysis of panel data which I did under the auspices of the Center for Advanced Study in the Behavioral Sciences (see Wiggins, 1955b), based on data supplied by Paul Lazarsfeld from a panel study done by a commercial research firm sponsored by a consortium of automobile companies, Lazarsfeld had developed the concept of "nearness to purchase" as a basic dependent variable. It was found that many respondents, over the three panel waves, moved further from purchase, rather than towards it, without any seemingly good substantive reason. Here again, without unravelling the different kinds of changes, it is almost impossible to understand the manifest findings. Once the changes resulting from the latent—manifest relationship are removed, consumers will be found fairly regularly moving towards purchase over time (of course, there are financial accidents and so on, latent changes of plans, etc., so that we never get total consistency or regularity).

It is very important to keep in mind that the element of probability involved in the latent—manifest relationship, while that portion of it which is based on error of measurement may and should be reduced within practicable cost limits as much as possible, is never going to be eliminated from most of the kinds of human behavior and attitude we are dealing with here. In the early stages of an election campaign, in selecting a brand of coffee from the store shelf, in deciding whether the government should allocate X or Y percent of its budget to a new educational program, and so on ad infinitum, a great many people are never going to know and act and report consistently and definitively their ultimate or basic positions. They may be forced to act in a certain manner at a certain time, as on election day, but even as they push the lever there is an element of "crass casualty." Market research workers speak of "impulse buying," which partakes heavily of this element. Anyone who has ever had to buy a new car in a big hurry, as I have had to do twice, once in the United States and once in Germany, knows that even a large purchase can incorporate a very large random component.

It was inevitable that some smart young computer software

expert who had taken elementary accounting and was familiar with the cost accountant's acronyms FIFO and LIFO would come up early on with the term GIGO — "Garbage In, Garbage Out". Of course, if you are studying educational achievement and you put the data for one class into the computer and think it is the data for a different class, you are in a GIGO situation. But the essence of the panel data situation is that what was loosely referred to as "garbage" in Chapter 1 above is for the purposes of the study of change basically a probability which is inherent and important and cannot be removed, but can be resolved analytically, as a useful, necessary, and important component of the change process. If we wish to use the rough analogies of Sir Alexander Fleming and penicillin, or the Japanese production of building blocks out of garbage, or re-cycling in general, so much the better. We can never dispose of this element in the production of panel data, but without handling it analytically, we can never understand what is happening.

In one of his "Theses on Feuerbach" Marx said that "Philosophers in the past have discussed reality; the point, however, is to change it." Easier said than done. In this book I have said little in the way of criticism of wrong interpretations made by many analysts (including myself) of change over time by using panel data. My view is that a substantial proportion of the inferences reported are erroneous. Nevertheless, I have always taken the view that the important thing is not to criticize, but to try to solve the problem. I do not deprecate the criticisms of others, often valid, often useful, but I place them secondary to positive efforts to deal with the problems.

Latent Probability Models and Social Theory

It may seem that there is an ad hoc character to some of the applications which have been made, and that is not totally without truth. The rigors enforced by small numbers of cases, in some instances, or by the necessity for having enough independent items of information for a mathematical solution to be possible, or the esthetic and practical pressure of Occam's razor, forcing a Hobson's choice between parsimony and realism — all these things have conspired to produce a seeming arbitrariness in some examples, which have indeed been offered occasionally primarily as examples.

However, there are substantial mathematical, empirical, and theoretical uniformities which have emerged. It will be recalled from Chapter 1 that (omitting the question of measurement space) some nine classes of models were described, based on the combination of three classes each of movements of the latent probabilities and the latent proportions, as follows:

(1) The latent probabilities
  (a) No change over time
  (b) Other systematic change over time
  (c) Unsystematic change over time
(2) The latent proportions
  (a) No change over time
  (b) Other systematic change over time
  (c) Unsystematic change over time

In using these models for various sets of non-experimental (in nearly all examples) data, certain empirical regularities have emerged, and these are related to the structure of the situation, the content of the variable, and they can be connected in various ways by several bodies of social theory.

In addition, a small paradigm of turnover for a three wave panel was shown in the first chapter, which noted that the amount of turnover (e.g., the total amount of individual change in both directions for a dichotomous item) could be equal for the time interval from time 1 to time 2 and the time interval from time 1 to time 3; it could be greater for the longer interval; it could be smaller or equal for the interval from time 2 to time 3 than for the interval from time 1 to time 2; and of course there are many other possibilities. Those relationships, which of course play a major part in determining the model parameters for a particular example, are also related to the structure of the situation, the item content, and to certain bodies of theory.

Once we get past the notion that what we are talking about is nothing but measurement error — a necessary but fuzzy concept in this field at best — we can, as we must, drop the idea that the measurement space or type of measurement must be the same or similar on the manifest and latent levels. It may be the same or it may not, and the fact is that it generally is not. Indeed, there may be no position at all on the latent level. This could be and has at times been called a zero reliability, but it is much more productive to think in terms of the fact that the respondent or subject simply has no position on the matter in question. Here we tend to be confused by physical analogies. Everybody has a weight and height, for example, but not everyone has an opinion on every political issue.

We are speaking here of the relationship of certain kinds of theory to the models and data under discussion. The term "theory" is perhaps too strong for what we are talking about, and is used loosely here. We do have certain concepts, and these lead to certain qualitative propositions which relate to some of the types of models, but the range of subject matter covered is so broad that clearly there is no embracing theory covering them all, without consideration of content.

In terms of the models, we are dealing with two basic concepts — the latent probability and the latent position. The latter depends on a host of substantive considerations in any particular instance of behavior, and changes in the latent position also depend on such a variety of substantive considerations, that little can be said of a general nature about them. Much more can be said about the latent probabilities, and it is to these that we shall primarily address this section.

Also, while it has been noted that with appropriate modifications the approach taken here can be applied to social aggregates, such as voting records of counties, retail store audits, social institutions of various kinds, the concentration here has been primarily on the level of the individual and even the single item of individual behavior or response, so that the concepts referred to will be primarily on that level.

We can first ask most profitably, "under what conditions is the latent probability high or low?" We can consider that problem at one point in time, and with some concepts in mind we can then consider the conditions under which the latent probability would be expected to remain stable over time, and under what conditions it would rise or fall over time. We can then classify the examples used in various models and offer some explanation for the reasons why certain models were appropriate in some situations as against other possibilities.

The basic notion of a distinction between the latent and the manifest is of course rooted in and related to many concepts in many fields. In biology the genotype—phenotype distinction is obviously a related notion; this has also been used a great deal by some psychologists. Pareto's concepts of sentiments, residues, and derivations also has some relationship to this concept, even though his own formulation has many specific features which do not need to be invoked here.

Basically, Pareto was more interested in the relation of large theories to the facts upon which they were supposed to be based, than to the relation between latent concepts and manifest positions. He was a grand theorist of the style which is not as popular today as it was fifty years ago in his field. He developed elaborate schema and a somewhat esoteric terminology which is not widely used today. Nevertheless, some of his concepts are pertinent to the thinking here.

Pareto was keenly aware of the difference between observables and non-observables in the study of human society and the human mind and behavior, and while he strongly advocated the use of observables to reach the non-observables, he was particularly incensed about what he considered to be conclusions not based on observables. Perhaps the most succinct statement of his views is to be found in the Appendix to his Treatise entitled "Index-Summary of Theorems". He divides "the facts that are observable in human societies" into two categories, of which the first is of interest here. This is described as "Manifestations, verbal or through conduct, of instincts, sentiments, inclinations, appetites, interests, etc., and the logical or pseudo-logical inferences that are drawn from such manifestations." This comprises both logical and non-logical conduct, and the latter is of particular interest in this monograph. He further divides the non-logical category into a part "that does not give rise to verbal manifestations" and a part that does, which he calls *c*.

He says that "The element *c* is outstanding in human beings, for they are wont to express their instincts, sentiments, etc. in verbal form .... The element *c* is divisible into two further parts:

(a) A part that varies but slightly (residues)

(b) A part that is much more variable (derivations)."

Thus a given manifestation may be divided into a relatively fixed residue which may give rise to a number of different derivations. This relationship is generally probabilistic if what Pareto calls the "logico-experimental" approach (which in his terminology includes the approach used here) is taken.

While Pareto was more interested in inferences in the form of propositions than in individual concepts, it is of course true that propositions include concepts as well as statements of their relationships, so the line of thought taken in this study can be handily fitted into that part of his theory which has been discussed. This does not imply acceptance of all the particulars of his elaborate and fascinating theories.

More directly relevant, perhaps, are some of the social psychological concepts of Kurt Lewin, and others still more recent. Lewin developed several concepts which are useful in thinking about the implications of the present models. Among these we may include the notion of "levels of reality or irreality". It is worth quoting Lewin (1935) on this point:

> The psychological environment of the adult shows a rather marked differentiation into strata of various degrees of reality. The plane of reality may be characterized briefly as the plane of 'facts' to which an *existence independent of the individual's own wishes* is ascribed. It is the realm of realistic behavior, of insuperable difficulties, etc.
>
> The more unreal planes are those of hopes and dreams, often of ideology. A stratum of greater unreality is dynamically characterized as a more fluid medium. Limits and barriers in such a stratum are less firm. The boundary between the self and the environment is also more fluid. In a plane of unreality 'one can do what he pleases'.
>
> A complete description of the psychological environment must always set forth the structure not only of the level of reality but also of the levels of unreality. If conditions on the plane of reality become too disagreeable for any reason, for example, as a result of too high tension, there arises a strong tendency to go out of the level of reality into one of unreality (flight into dream, into phantasy, or even into illness).

We do not have to subscribe to all of Lewin's "topological representations" to recognize that the notion of a high degree of fluidity across boundaries where the level of reality is low corresponds in many ways to a low latent probability. While not every phantasy or ideology is fluid — some may indeed represent fixations, obsessions, or other relatively unchangeable notions, in part — yet it is reasonable to think of a spectrum of attitudes and beliefs which range from the relatively firm or "hard" at one end to the relatively "soft" at the other.

There are many cognate concepts in social psychology and related fields which relate to the firmness with which an idea or sentiment is held. Some of these may be listed as follows:

(1) *Intensity* of a conviction. Generally, the more intensely a conviction is held the less labile it is, and the higher the latent probability.

(2) *Commitment.* The more highly committed a person is to a given view, the less likely he is to change. Thus if a voter has given a contribution in time or money to a candidate, he has become more highly committed to that candidate, and is less likely to change.

(3) *Saliency.* If an issue is not in the forefront of consciousness,

it is not likely to have the consistency which another issue very much in the front of consciousness has. A person who has never thought about or rarely thought about the use of fluorides in water supplies is likely to have a lower latent probability in responding on this issue than a person who has been actively spending a good deal of time thinking about the issue.

(4) *Extremity* of a position. Generally, as has been shown many times, extreme positions are held more firmly than middle positions.

(5) *Degree of structurization of a situation.* A highly structured situation is likely to yield greater stability or consistency in behavior than a weakly structured situation.

(6) *Involvement.* Some writers view this as essentially the equivalent of commitment; others as somewhat different. Whatever the precise definition taken, a deeper involvement usually implies a more consistent behavior, thus a higher latent probability.

(7) *Consistency.* If we get into the relationship of different items of attitude or behavior, we generally find that unambiguous items which do not conflict with others are more stable and would therefore show higher latent probabilities than those involving conflicting views, as are often introduced in terms of the concepts of cross-pressures, dissonance, and the like.

(8) *Information.* In general, the greater the depth of information upon which an opinion or act is based, the greater will be the latent probability.

These general observations hold whatever specific view we take of a particular theory, or even if they are looked at from the point of view of simple common sense. There is a good discussion of some of these notions in Festinger (1957) and Kiesler *et al.* (1969). Other related notions such as Festinger's (1957) definition of "confidence" could also have been introduced here.

There is again another line of related developments arising from studies of learning in both animals and children. We often read here that learning starts with "random behavior" or "trial and error". This means low latent probability, in our terms. Whether a child hits the second or third slat of his cradle with his rattle at a certain time on a certain day is in a sense random. This does not mean that it is undetermined, but that we are not interested in studying his maturation or learning in the kinds of minutiae which determine such individual outcomes. The fact that there is a random component from the point of view of a given level of explanation does not mean there is not a systematic process of maturation or learning going on at the same time. It is a question of dividing the behavior into latent position and random behavior, the latter of which determines the latent probability.

All of these comments have borne on the reasons why latent probabilities may be high or low at a given point in time, but they also give insight into possible changes in latent probabilities over time, since the concepts with which we are dealing refer to matters which may change over time, or which may not change over time. We can look at some of the conditions under which they do or do not change.

One of the complexities which arises in talking about this topic is that, whereas in general latent probabilities are indifferently related to the time interval used for measurement, the latent positions may be highly related to the time interval, if they are changing. An analogy may be drawn to measurements of human height, where if they are taken at six-month intervals when the subjects are one year old, two years old, etc., there may be large proportionate changes on the latent level. If, on the other hand, measurements of human height are taken at the age of thirty, over a similar interval, there is little change on the latent level.

Again, when we study gambling, if we are dealing with accurately made and operated devices, the latent probability of a coin falling heads or tails, of a given poker hand, of red versus black in roulette, and so on, does not change at all, regardless of the time interval (assuming the devices do not wear down and the game is "honest").

Or, in meteorology, it is possible to pick a sample of spots on the earth's surface, such as rain forests and deserts, where the latent position is almost constant.

In this case, if accurate measurements are made, the turnover would be virtually zero, whereas in the case of roulette it would be extremely high, the sole reason being latent probability, with no change in latent position in either case.

On the other hand, in temperate climates, systematic average changes in temperature and rainfall occur and are recorded continually in many places. Some of the larger processes are well understood and highly predictable, such as the relationship of the earth to the sun, and they yield highly reliable average estimates, such as that in the Northern Hemisphere in many places July 1 will be hotter than January 1 in any given year.

For detailed prediction at a given place, say whether it will be raining in Manhattan's Central Park, a time interval of one minute yields a very reliable forecast. When the meteorologist tries to predict rainfall one week ahead, or even one day, the reliability of his forecasts drops enormously. The larger systematic changes over the year which are fairly well understood are too large and too long in occurring, and the short-run experience forecast breaks down — the weather changes. Whether it is going to rain at 1:00 A.M. next Thursday week depends on a host of minutiae — clouds, wind velocity and direction at many altitudes, temperature and humidity at many altitudes — which are so numerous and difficult to measure a week in advance that they amount to a random variable, in the aggregate.

Thus we are dealing here with a systematic change in latent positions, plus what amounts to a random variable, which itself may change according to season and other factors. This corresponds to a low and changing latent probability.

The paradigm which was given for a three wave panel taken in pairs of waves in Chapter 1 gave the following picture of a two wave table:

		Time 1	
		Yes	No
Time 2	Yes	A	B
	No	C	D

The sum of $B + C$ was called the "turnover", and represented for a dichotomous attribute the total number of those who changed in either direction. For games of pure chance, using, say, a sample of successive pairs of outcomes in coin tossing, without any change in latent position, this number would (disregarding as we always do sampling variations) equal the sum $A + D$. In other words, it would show an enormous turnover, although on the latent level nothing had changed. Of course, classical probability theory (not only the interest of French nobles and Pascal, but also Edgar Allen Poe, Arthur Conan Doyle, and innumerable gamblers and suspense story writers), usually assumed the independence of one event or time point from the succeeding event or time point — the contrary of Markov process analysis, where the succeeding events are dependent, even if only on one time point — whereas we are here arguing that in most human behavior and response there is an admixture of the two kinds of relationships. There is a random element which is not dependent on the preceding position, and an element which is. Without separating these analytically, we cannot understand what is happening.

There are processes which do not involve learning, maturation, or any extrinsic element except the intervention of events which are entirely accidental and random on the level of analysis which is being attempted. Stimulus-response theories are of no particular value in many of these situations, which are non-experimental in nature, nor are neo-Kantian theories of innate, universal, or maturational processes, which may be very useful in other contexts. We do not therefore always need to invoke philosophy, structural linguistics, stimulus-response theories, or other theories of like sort which attempt to explain the relationship of behavior or attitude at time 1 as related to time 2, any more than we do in a roulette game. On the other hand, such processes may be occurring simultaneously with the kind of non-incorporated probability elements of which we are speaking.

If there is no process occurring other than that which occurs at Las Vegas or Monte Carlo, we may have high or low turnover according to whether the relationship between time 1 and time 2 is based on a low or high latent probability. Thus in gaming on even odds the turnover would be quite high, while the turnover of human height for subjects aged about thirty would be relatively quite low. In the attitude—opinion field, we might call the latter type of situation a *central* process, and the former might be referred to as a *peripheral* process, based on the concepts we have invoked earlier. The definition, to be sure, is weak in the sense that it is operational, but the concepts used provide some theoretical basis for the distinction.

It will be recalled that in the paradigm of Chapter 1 we depicted the pairwise relationships of times 1 and 3 and times 2 and 3 as follows:

		Time 1	
		Yes	No
Time 3	Yes	$A-\alpha$	$B+\alpha$
	No	$C+\alpha$	$D-\alpha$

		Time 2	
		Yes	No
Time 3	Yes	$A+\beta$	$B-\beta$
	No	$C-\beta$	$D+\beta$

If the latent positions are not changing at all, and the item being studied is one of high latent probability which itself is not changing, we have low turnover from time 1 to time 2. We also have the fact that α and β are zero — that is, the relationship between time 1 and time 3 is the same as that between time 1 and time 2, and is the same as the relationship between time 2 and time 3. If everything is the same except that the item studied is one of low latent probability which itself is not changing, we have high turnover for each pairwise relationship, but it is the same for each of the three pairwise relationships. When the central processes meet these other conditions, they will show low but constant turnover, whereas when the peripheral processes meet these other conditions they will show high but constant turnover.

If there is a process of change in latent positions going on which itself produces constant change over a given period of time, then these changes are added to the changes which appear as a result of latent probability. Some latent processes have this character and are well understood; most do not, certainly in the attitude—opinion and consumer behavior fields. If they do, however, the relationship between time 2 and time 3 will be similar to that between time 1 and time 2, but the relationship between time 1 and time 3 will show a larger turnover; that is, β will be zero but α will be a positive number. That will be true even though we are dealing with a process already in equilibrium, but the total size of the turnover will depend not only on the amount of change occurring in the latent positions but also on the level of the latent probabilities, with the low probabilities again producing the higher turnover. A stable equilibrium process among the latent positions may be a peripheral process with low latent probabilities, thus giving rise to relatively high manifest turnover.

Obviously, in choosing a model for an attempt at fitting, the more we know or can conjecture on grounds of previous research, substantively and theoretically, the better we can decide what kind of process we may be dealing with in terms of the latent positions and the latent probabilities as well. If in terms used above we have reason to think that there is no change of latent position or of latent probability, the earliest and simplest models of this monograph will be useful. As it happens, attention usually is directed to processes where some changes of one or both types of parameters are expected, and while we may have a qualitative sense for what those changes might be, we rarely have a precise notion of what they will be. For example, if a new brand of a product is introduced, we generally expect that it will get some share of the market; thus its latent proportion will increase, while that of other brands will decrease. In quasi-experiments, an effect in one direction is often predicted. On the other hand, political campaigns are notoriously full of surprises, although sometimes there may be a basis for thinking that over time a certain direction of movement will occur.

Generally, if the process we are studying is one in which the basis for some or all of the concepts we have mentioned is changing in some systematic way, we may expect the latent probabilities to change accordingly. We usually have a much stronger basis for this kind of prediction than for a prediction about latent positions. For example, a useful proposition is Lewin's notion that in locomotion to a goal the level of reality generally increases. Some of the processes we study are simply *continuing*, meaning that they do not move to a terminus and if studied after they have been in operation for a fair period of time should not bring about a change in latent probabilities.

Other processes we study may be called *terminal*, meaning that they have a beginning, a middle, and an end, as in a political campaign. Of course, the respondents have positions after the election, so certain items including many issue items do not really come to an end. Usually, however, after the actual election a substantial proportion of the items related to the campaign lose saliency, involvement, and so on. This is also true of a great many quasi-experiments, or of a school year, and many other situations which are being studied by means of panels.

Whereas in the continuing processes we do not expect the latent probabilities to change, in the terminal processes, as they near their terminus or goal, we may expect the latent probabilities to rise. Ceteris paribus, this is what we have actually found in nearly all examples of this monograph which fall into this class of processes, which we shall list below. It is also worth noting that in Coleman's (1964) quite different examples the same result is noted for his continuous-time models, where the data are examined in a manner which permits a conclusion on this point.

Another related proposition is the frustration—regression phenomenon. With both animals and human subjects, it is often found experimentally that under frustration there is, among other things, increasing randomness of behavior; i.e., a lower latent probability. H. Goldhamer once suggested to me a series of experiments in which this hypothesis could be tested. Although I was never able to run these experiments, it is a matter of common observation and introspection that under frustration one is likely to get "rattled", that is, behavior or opinions may become more random in character. This does not eliminate a purposive element, or an element of learning, extinction, maturation and the like. When we say that a person "regresses" to the level of activity appropriate to a less developed state, we do not mean that the lower level has nothing but a random character. Generally, however, and even if we agree with some of the neo-Kantian or Wordsworthian notions now becoming popular, it is a fact that the movements of a baby with his rattle incorporate a larger random element than those of, let us say, a good secretary. The terms "rattle" and "getting rattled" have an infinity.

Thus even though the baby may be learning in the course of his seemingly random movements, the idea of regression involves increasing randomness or a decrement of learned and purposive behavior. Consequently, it should in terms of the present models imply a lower latent probability.

This thought can bring us back to the distinction between "learning" and "non-learning" models, and it implies that most processes studied by panels may well partake of both. The emphasis here is on the latter, the part of behavior which is not carried over from one time to the other, as distinct from the variants of Markovian models which emphasize the part which is carried over. To the extent that "non-learning" is involved, the latter models will generally not fit as well as they might; the converse, of course, may also be true. However, if both are involved, and the model provides for both, then if sufficient data are available in a given problem good fits should be found. To echo Coleman again, they are not likely to be parsimonious.

If we are studying processes in which there is an increase in latent probabilities but no change in latent positions, we would have the seeming paradox that the relationship between manifest positions at time 1 and time 3 would be higher than those between time 1 and time 2, while the relationship between time 2 and time 3 would be the highest. In the study of numerous examples, we do not usually get this result. Generally, we find that the relationship is highest between time 2 and time 3, but the relationship between time 1 and time 2 is higher than that between time 1 and time 3. This leads to the conclusion that in most actual terminal processes we are getting a combination of increasing latent probabilities and some change of latent positions. The latter changes, which increase with time, are reducing the relationship between the manifest positions at time 1 and time 3, even though on the basis of increase of latent probabilities alone the relationship would be expected to increase. Hence merely looking at the pairwise turnover tables alone can be very misleading unless we employ a model with regard to the latent probabilities.

However, as noted, any adventitious influence can serve to explain changes in latent positions, although of course we are always looking for mathematical or theoretical models. Changes in the latent probabilities alone, such as their rising as a respondent moves towards a terminus in a given process, can account for such phenomena as increasing correlations in a system if variables, or increasing polarization between two positions on an issue or candidate. Such movements do not require any substantive theory, beyond the various concepts and theories which account for the changes in latent probabilities. To be specific, we do not require such notions as "cognitive dissonance", "learning", "maturation", "polarization", and so on. In that sense we may have gained a degree of parsimony in non-experimental studies, as well as a degree of insight into the substantive situation.

We may end this essay by showing how the concepts, models, and examples fit together, in a qualitative way. To do this we shall display a rough division of the way data and models should look, and the way examples do look, under a number of rubrics. Some which have been discussed will be repeated very briefly in this summary. There are five kinds of things we wish to mention here, which are (1) the type of item or behavior we are dealing with, (2) the question of high or low latent probability, (3) the type of process we are dealing with when time is brought into consideration, (4) the types of models which are appropriate for various types of processes, and (5) examples, both those taken from empirical studies reported in this monograph and a few hypothetical examples, showing how various types of models yielded certain results or might yield certain results.

This discussion will be brief and rather loose, since we cover in differing degrees so many types of behavior and so many models. Thus we shall talk of "related concepts", a vague term at best, with the aim of giving some overview of the ramifications of the present approach, without pretending to a specificity which appears only in parts of the actual work on models and data.

(1) Item Characteristics

Here we shall merely divide items of response or behavior into "hard" and "soft", recognizing that this is a crude dichotomization which covers a multitude of variations and types of behavior. In some of the terms we have been using we would have the following kinds of related concepts:

"Hard"
Residues (Pareto)
High level of reality (Lewin)
High level of learning
High intensity
High saliency
High commitment
High confidence
High involvement
Extreme position
High structurization of a situation
Impermeable boundaries (Lewin)
Lack of fluidity of movement through barriers (Lewin)
High information
Central (versus peripheral) attitude or trait
Factual data (versus opinion)

All of these are on the "hard" side, and their converse is on the "soft" side. For example, we would have "derivations" (Pareto), "high level of irreality" (Lewin), "permeable boundaries" (Lewin), "fluid movement through barriers" (Lewin), and so on, on the "soft" side. Some refer more to the subject, some more to the situation. Many other related notions could be introduced — for example, in marketing the purchase of a large ticket item is generally "harder" than that of a very small ticket item, though not always so. Of course, it is not meant to imply that all these characteristics have the same meaning, but for the present broad purpose they are sufficiently related though diverse to give some indication of what is meant here by the simple division of item characteristics into "hard" and "soft".

(2) High or Low Latent Probability

Terms such as randomness, accident, error, casualty, trial-and-error, and the like are related to the notion of probability — when they are high the latent probability is low. Of course, probability is a continuous variable, but for this summary it is sufficient to divide it into "high" and "low".

The general argument we have made is that hard items are associated with high latent probabilities, ceteris paribus, and soft items with low latent probabilities.

(3) Types of Processes Over Time

Even if an item is quite "hard", we may get a process of change occurring in manifest data over time. More evidently, there are a great many processes which clearly do permit or require change over time, in terms of the characteristics of the sort listed above. Thus in a political campaign, or in a school year of education, or when a new brand or product is marketed, or in a quasi-experiment, and so on, we expect to find changes occurring. Sometimes we have a very well defined expectation about the changes that will occur; often we do not. We may have experience or theory or both as a guide, or we may not.

Among the notions which have seemed useful in classifying processes are the idea of locomotion to a goal (Lewin), continuing versus terminal processes (in a terminal process, the campaign ends and the election is held, or the school year ends, or someone gets married, etc.), central versus peripheral processes, and so on. A process which we may call terminal does not necessarily have to end forever; it may only reach an intermediate stopping point and resume at a later point.

Leaving aside the question of changes in latent position, time may enter in an important way into thinking about what happens to the latent probabilities. In a continuing process, where none of the variables which determine the level of latent process may be expected to change, we do not expect the latent probabilities to change. In a terminal process, on the other hand, the general situation is that the closer in time the terminus is to come, the higher the level of reality, and in general the higher are the variables associated with higher latent probabilities. Samuel Johnson said that "when a man knows he is going to be hanged in one month, it concentrates his mind wonderfully," a remark Winston Churchill liked to repeat. Lewin says that as ego approaches its goal the level of reality is raised. Certainly, for most students the final examination at the end of a course assumes much greater reality and involvement during the final week of a term than during the first week.

The converse is also generally true. After an election interest and involvement usually decline, or during the summer a student's interest and involvement in his achievement usually decline. Thus the relevant variables may change either way, depending on the movement of the situation. We may, therefore, expect the latent probabilities to move accordingly. Of course, the matter is not always so simple. There may be processes in which interventions, deliberately introduced or accidental, have the effect of shaking confidence, while at the same time they may increase saliency. This is particularly true of surprising or unexpected events which may occur. Thus in field research particularly some of the relevant variables may be moving at cross purposes or in opposite directions to each other. That of course complicates the process greatly, and may make it impossible to form a reasonable expectation as to the total outcome.

Beyond this, we may or may not have a reasonable expectation that the latent positions will not change, or will change in one direction to some unspecified degree, or will change in both directions to an unspecified degree, or will change in some systematic way. That of course will affect the appearance of the manifest data. These changes, compounded with those of latent probabilities, may make the manifest turnover an extremely dangerous indicator of the process, if taken at face value.

(4) Types of Processes and Types of Models

In principle, as we have noted, latent process models can be extended to any number of time points (or panel waves), any type of latent distribution or manifest measurement, any number of categories of a non-metric sort, and any of a great number of assumptions about the relationships required for development of the algorithms. Further, almost any model of temporal process (contagion, diffusion, etc.) on the manifest level is a fit candidate for an analogous model on the latent level. In addition, mathematical models could be developed for changes in the latent probabilities in cases where sufficient theory and experiment were available.

However, the complexity mounts very rapidly, and while we have shown a few of these possibilities, they are best reserved for attacks on specific problems by detailed examination rather than for a general discussion such as this. We have therefore tended to use chiefly models of dichotomous data with only three time points, the minimum from which in most instances any useful information can be derived. This has required rather severe assumptions, and it may be surprising that we have achieved as good results as we have.

The situation is not too different from that in ordinary multivariate analysis, where limitations of sample size, cost, etc., usually force the analyst to use fewer categories than he would like, fewer variables, and so on. The same strictures apply to most simulation work.

For much the same reasons, any attempt to display all the possible cross-relations of the factors we have introduced here as related to these models has been abandoned. It would lead to a very unwieldy display — among other things, a typographic nightmare. Instead, we have chosen, as in the text above, to single out certain promising points, and often to speak of them in a qualitative or suggestive way.

We have noted that the models here are based on two types of parameters, either or both of which can change over time in various ways. Thus we have the latent positions and the latent probabilities, either of which over time can display no change, other systematic change, or unsystematic change in one or more directions. If we cross these possibilities, we have a very large number of joint possibilities, of which we have considered a few. By listing these possibilities, confining the terminology to such relations as "high" and "low" or "changing or unchanging over time", we can briefly discuss the kinds of models which are likely to be useful in the analysis of certain kinds of substantive situations.

The simplest situations are those in which neither latent position nor latent probability changes in magnitude. If the latent probabilities are low, as discussed above, the turnover in the manifest data will tend to be high, and conversely. In most public-opinion panels the turnover over a pair of panel waves will be on the order of five to forty per cent of the total number of cases. The peripheral processes will generally show more change than the central. Further, the processes which are likely to fit this kind of model are generally stable, non-terminal in character. For example, a case which might fit, with fairly high turnover, would be exposure to a small repeated advertisement in a periodical of general circulation appearing at regular intervals, where the total circulation of the magazine is not itself undergoing any change. A special-purpose magazine advertising exposure might fit a similar model but with somewhat smaller turnover, based on a higher latent probability.

Changes in latent position are difficult to deal with on such a general basis, because so many extraneous disturbances or interventions might bring about substantial change over any given time period. Witness the 1972 presidential campaign in the United States. Sometimes we have reason to think that certain disturbances or interventions might push subjects in a given direction, in which case the models providing for no change in latent probability but with change in a particular direction on the latent level might occur. It is rare in field work to be able to assign a numerical value or predicted value to such a change in advance. Sometimes we are dealing with a definite long-term movement in one direction, which is known. Whether such qualitative knowledge can be put into mathematical terms is another matter. If so, we could have the model with no change in the latent probabilities but a systematic change in latent positions. In a few instances we might even have a systematic change in both directions on the latent positions, as in the latent Markov chain, without any change in the latent probabilities. Generally, attempts to locate such examples have met with poor success. In non-experimental field panels, certainly in the attitude—opinion field, it is usually the case that the magnitude of the extraneous disturbances simply over-rides any systematicity of movement among the latent positions.

When we come to consider changes in the latent probabilities, we are in a somewhat better position, although still on a qualitative level. Here we have much more general theory as a guide. Thus if a process involves movement towards a terminus or goal, we can generally predict that the latent probabilities will tend to rise, while once the terminus is past they will tend to decline. We still are rarely in a position to put numbers on the changes in magnitude of the latent probabilities. Since we often deal with situations in which we know some of the characteristics of the process in a qualitative way, we can often fit models providing for movements of the latent probabilities. Of course, the magnitude of the turnover will be greater in general the further we are in time from the terminus, ceteris paribus, and it may even be swamped by changes in the latent positions. Usually, however, and particularly in peripheral items, matters of marginal interest to the subject, regressive behavior, or when other variables mentioned previously are in the appropriate position, the latent probabilities will account for a substantial amount of the turnover.

Thus the type of model we choose to attempt to fit will depend on how much we know about the item in question, in terms of the structure of the situation and previous research and theory. The same thing applies, of course, to the assumptions we wish to make about independence over time, the number of categories used in attribute models, and so on. In the absence of any knowledge, or the presence of almost no knowledge — often the case — we may apply a model on a more or less ad hoc basis, hoping to learn something inductively about the case at hand and something which will aid in the continuing process of future research.

(5) Some of the Examples Given

Not all the models attempted have been reported, and not all models reported showed good fits to the data. In some instances, alternative models might have fit better. On the other hand, there have been some remarkably good fits to fairly simple models. Sometimes there was an ad hoc character to the selection of a model, other times a good theoretical basis for choosing. There is not a great deal to be gained, therefore, by trying to report all the particular examples and fits of particular selected items to particular models.

A partial summary, however, might be helpful as it may demonstrate the range and power of the general approach taken across a wide variety of situations, as well as provide a basis for the generalizations which might be tentatively made. For this purpose we may class the models into four groups according to whether the latent probabilities and latent positions were unchanging or changing, with notes on the types of changes where they occurred and on the substantive types of situations.

(a) No Change in Latent Probabilities, No Change in Latent Positions

Item	*Situation*
Education, 1940 voting study	Continuing
Attitude towards length of war, WW2, control group	Continuing

(b) No Change in Latent Probabilities, Change in Latent Positions

Item	*Situation*
Attitude towards length of war, WW2, experimental group	Terminal (experimental)
Magazine readership, two periodicals	Continuing with changing circulation
Winner expectations in 1940 voting study	Poor fit, terminal but in flux

(c) Change in Latent Probabilities, No Change of Latent Position

Item	*Situation*
Grades of medical students over three years	Terminal (latent probabilities rose)

(d) Change in Latent Probabilities, Change of Latent Position

Item	*Situation*
Vote intention, 1948 study	Terminal (latent probabilities rose)
Strength of feeling about political choice, 1948	Terminal (latent probabilities rose)
Interest in political campaign, 1948	Terminal (latent probabilities rose)
Source of information about campaign, 1940	Terminal (latent probabilities rose)
Parents yell at children too much, Project ENABLE	Quasi-experiment, terminus at time 2, latent probabilities up then down, latent position moved in one direction
Is parent willing for child to be entertainer? Project ENABLE	Latent probabilities went up, latent position moved in one direction
Knowledge of community resources, Project ENABLE	Latent probabilities went up, latent position moved in one direction
Do children come to parent with problems? Project ENABLE	Latent probabilities went up, latent position moved in one direction. All Project ENABLE situations were quasi-experimental between time 1 and time 2, with possible sleeper effects

Generalizations and Conclusion

(1) Any measured observation of human behavior contains a random element which is *not* incorporated into subsequent behavior. The variable or item under investigation is called the "latent"

variable; the observation which goes into the computer is called the "manifest" variable. The presence of the random element makes the relationship between the two probabilistic. However, this does not require any particular philosophical theory of determinism or non-determinism, of axiology, or of causation or non-causation. At a minimum, there is always some error of measurement, but the random element usually is far larger than would be explained by what is ordinarily meant by error of measurement or unreliability. Often error of measurement can be reduced to a small amount if enough expense and care is incurred; alternatively, if, let us say, the investigator is interested in errors in key-punching, he can predict the probability of a given type of error if he finds out enough about the capabilities and experience of the operators; the age, quality, and capabilities of the punch machines; the conditions of work, such as lighting, time pressure, supervision, etc.; and a host of other minutiae which are not objects of the primary investigation. Usually this is not worth the expense, and the resulting error is simply treated as a random variable.

There is much substantive theory to confirm the fact that even when such errors of measurement are removed a component of randomness remains in the manifest data. This does not preclude the existence of other random elements which may become incorporated into subsequent behavior, as in learning situations. They would be incorporated into the latent position, while the random element which is not carried over would be incorporated only into the manifest position.

Of course, the approach taken here does conflict with certain alternative models for the same behavior, and subsequent investigations will no doubt resolve these issues more definitively.

(2) We refer to the relationship between the latent position and the manifest position as the "latent probability". This probability varies enormously in different situations, under different conditions, as outlined above. It can easily account, in a panel item which is an observed set of categories, for changes amounting to as much as forty or more percent of the total number of observations.

It is important to keep in mind that the form or metric of the variable on the latent level does not necessarily have any relation to the form or type of measurement on the manifest level. It is evident that a continuous variable on the latent level can easily be observed or manipulated into a "lower" form of measurement on the manifest level (e.g., an interval scale into a manifest dichotomy), and this is commonly done in field work. The converse also happens, but not as frequently. For example, the number of children born to a given man and woman is a fairly small integer, highly skewed in form of distribution with most samples, yet it has occasionally been treated as a continuous variable. This kind of usage has tended to go out of fashion, but not entirely. Incidentally, this is a good example of the previous point, for with care and expense the error of measurement can be virtually eliminated from measurements of the number of children, whereas there is a very large random element in the behavior itself in a great many cases; that is, the number of children actually born to a given couple contains very often a large random component.

Again, psychologists will build up something which on the manifest level approximates a continuous variable by summing a large number of dichotomous responses, and then often turn around and treat the result as a small set of ordered categories. Clearly, there is a very limited inherent relationship in the form, and that fact would almost invariably yield a probabilistic relationship between latent and manifest.

(3) The magnitude of the latent probability will vary greatly from one item to another, and will very often be so great in field studies that it accounts for much more of the change in the manifest data than do changes in the latent position, the variable under investigation. This can and does lead to a host of erroneous interpretations of the manifest changes, based on the assumption that they are in fact changes of the latent position. For this reason, a great mass of substantive panel analysis is suspect and with sufficient effort might be shown to be wrong. One difficulty is that the minimum three wave data which would usually be needed are not reported. Another is that it is more important to light a candle than curse the darkness.

(4) Further to complicate matters, the conditions giving rise to higher or lower latent probabilities often change during the time period studied, so that the latent probabilities change. This complicates the analysis, particularly in the absence of solid models for changes in the latent positions, which are usually lacking except in tightly controlled laboratory experiments.

(5) In a qualitative way, we have theories which will account for certain directions of change of magnitude on the part of the latent probabilities in many situations. No one of these theories is required, but they are generally compatible with the empirical findings where good fits were obtained. For example, we expect, and

find, in nearly all instances that the latent probabilities of an item will increase over the duration of a terminal process.

(6) There are evidently scores of directions in which the point of view advanced here could profitably move. William of Occam admonished us not to multiply entities beyond necessity, but even more perhaps than in modern physics we are going to find necessity a hard taskmaster. Still, in time, with our increasing technical capabilities in data handling, we may be able to produce some results (and in fact have already produced a few) which go beyond sheer speculation. It is hoped that the present approach will provide some leads in that direction.

APPENDIX A

An Alternative Derivation of the Solution to a Special Case

The case to be solved is that involving four interviews, two latent classes at any one time, with free movement between them through time, with latent probabilities unchanging through time, and with the terminal latent position independent of the initial latent position when the latent position at times 2 and 3 is fixed. The model is given in Chapter 3.

We have equations of the following types:

$$p_{1...} = a_1 v_{1...} + a_2 v_{2...}$$

$$p_{11..} = a_1^2 v_{11..} + a_1 a_2 (v_{12..} + v_{21..}) + a_2^2 v_{22..}$$

$$p_{111.} = a_1^3 v_{111.} + a_1^2 a_2 (v_{112.} + v_{121.} + v_{211.}) + a_1 a_2^2 (v_{122.} + v_{212.} + v_{221.}) + a_2^3 v_{222.}$$

$$p_{1111} = a_1^4 v_{1111} + a_1^3 a_2 (v_{1112} + v_{1121} + v_{1211} + v_{2111}) + a_1^2 a_2^2 (v_{1122} + v_{1212} + v_{1221} + v_{2112} + v_{2121} + v_{2211}) + a_1 a_2^3 (v_{1222} + v_{2122} + v_{2212} + v_{2221}) + a_2^4 v_{2222}$$

From equations of the first type we derive $v_{1...}$ etc. as follows:

$$p_{1...} = a_1 v_{1...} + a_2 (1 - v_{1...}) = a_2 + (a_1 - a_2) v_{1...}$$

$$v_{1...} = \frac{p_{1...} - a_2}{(a_1 - a_2)}$$

From equations of the second type we derive

$$p_{11..} = a_1^2 v_{11..} + a_1 a_2(v_{1...} - v_{11..} + v_{.1..} - v_{11..})$$
$$+ a_2^2(1 - v_{1...} - v_{.1..} + v_{11..})$$
$$= (a_1^2 - 2a_1a_2 + a_2^2)v_{11..} + (a_1a_2 - a_2^2)(v_{1...} + v_{.1..}) + a_2^2$$
$$= (a_1 - a_2)^2 v_{11..} + a_2(a_1 - a_2)(v_{1...} + v_{.1..}) + a_2^2$$

$$v_{11..} = \frac{p_{11..} - a_2(a_1 - a_2)(v_{1...} + v_{.1..}) - a_2^2}{(a_1 - a_2)^2}$$

Substituting for $v_{1...}$ and $v_{.1..}$ from equations of the first type gives

$$v_{11..} = \frac{p_{11..} - a_2(p_{1...} + p_{.1..}) + a_2^2}{(a_1 - a_2)^2}$$

Continuing this procedure through equations of the third and fourth types gives equations of the following kind:

$$v_{111.} = \frac{p_{111.} - a_2(p_{11..} + p_{1.1.} + p_{.11.}) + a_2^2(p_{1...} + p_{.1..} + p_{..1.}) - a_2^3}{(a_1 - a_2)^3}$$

$$v_{1111} = [p_{1111} - a_2(p_{111.} + p_{11.1} + p_{1.11} + p_{.111})$$
$$+ a_2^2(p_{11..} + p_{1.1.} + p_{1..1} + p_{.11.} + p_{.1.1} + p_{..11})$$
$$- a_2^3(p_{1...} + p_{.1..} + p_{..1.} + p_{...1}) + a_2^4]\left[\frac{1}{(a_1 - a_2)^4}\right]$$

Now the independence assumption implies the following:

$$\begin{vmatrix} v_{1111} & v_{1112} \\ v_{2111} & v_{2112} \end{vmatrix} = 0$$

Adding the first column of this determinant to the second, then adding the new first row to the second,

$$\begin{vmatrix} v_{1111} & v_{1112} \\ v_{2111} & v_{2112} \end{vmatrix} = \begin{vmatrix} v_{1111} & v_{111.} \\ v_{2111} & v_{211.} \end{vmatrix} = \begin{vmatrix} v_{1111} & v_{111.} \\ v_{.111} & v_{.11.} \end{vmatrix} = 0$$

or

$$v_{.11.} v_{1111} = v_{111.} v_{.111}$$

Looking at the equations in these unknowns on the previous page, we observe that the denominator of $v_{.11.}$ contains $(a_1-a_2)^2$ and the denominator of v_{1111} contains $(a_1-a_2)^4$, thus yielding for the denominator of the left hand side of the equation above $(a_1-a_2)^6$. $v_{111.}$ and $v_{.111}$ each contain $(a_1-a_2)^3$ in the denominator, thus yielding for the denominator of the right hand side of the equation above $(a_1-a_2)^6$. Multiplying the equation through by this term clears the denominator. The remaining equation has only one unknown, a_2. Multiplying out and collecting terms,

$$a_2^4(p_{1..1}-p_{1...}p_{...1})-a_2^3(p_{11.1}-p_{1...}p_{.1.1}+p_{1.11}-p_{1...}p_{..11}$$

$$+p_{.1..}p_{1..1}-p_{...1}p_{11..}+p_{..1.}p_{1.1.})+a_2^2(p_{1111}-p_{1...}p_{.111}$$

$$+p_{.1..}p_{11.1}-p_{11..}p_{.1.1}+p_{.1..}p_{1.11}-p_{1.1.}p_{.1.1}+p_{..1.}p_{11.1}$$

$$-p_{1.1.}p_{.1.1}+p_{..1.}p_{1.11}-p_{1.1.}p_{..11}+p_{.11.}p_{1..1}-p_{...1}p_{111.})$$

$$-a_2(p_{.1..}p_{1111}-p_{11..}p_{.111}+p_{..1.}p_{1111}-p_{1.1.}p_{.111}$$

$$+p_{.11.}p_{11.1}-p_{111.}p_{.1.1}+p_{.11.}p_{1.11}-p_{111.}p_{..11})$$

$$+(p_{.11.}p_{1111}-p_{111.}p_{.111})=0$$

In determinantal form, this is similar to equation (3.34) of Chapter 3. Now the equations for $p_{1...}$, $p_{11..}$, etc., with which we started, are symmetrical in a_1 and $v_{1...}$, $v_{11..}$, etc. on the one hand and a_2 and $v_{2...}$, $v_{22..}$, etc. on the other. If expressions were developed for $v_{2...}$, $v_{22..}$, etc., they would be identical with those for $v_{1...}$, $v_{11..}$, etc. except that a_1 and a_2 would be interchanged. For example

$$v_{2...}=\frac{p_{1...}-a_1}{(a_2-a_1)}$$

where we had

$$v_{1...}=\frac{p_{1...}-a_2}{(a_1-a_2)}$$

The independence assumption implies that

$$\begin{vmatrix} v_{1221} & v_{1222} \\ v_{2221} & v_{2222} \end{vmatrix}=0$$

Adding the second row of this determinant to the first, then adding the new second column to the first,

$$\begin{vmatrix} v_{1221} & v_{1222} \\ v_{2221} & v_{2222} \end{vmatrix} = \begin{vmatrix} v_{.221} & v_{.222} \\ v_{2221} & v_{2222} \end{vmatrix} = \begin{vmatrix} v_{.22.} & v_{.222} \\ v_{222.} & v_{2222} \end{vmatrix} = 0$$

or

$$v_{.22.}v_{2222} = v_{222.}v_{.222}$$

When the derived expressions in p and a are substituted in this equation, the denominator, containing a_2 and a_1, cancels out. The remaining equation is identical with the one obtained previously except that the argument is a_1 instead of a_2.

Hence a_1 and a_2 are roots of equation (3.34) of Chapter 3.

APPENDIX B

General Solution for Latent Probability Models for Items with Multiple Manifest Response Categories

The limitation of the latent probability models presented in the text to items with only two manifest response categories keeps the equations simple and the computations relatively economical, but is theoretically quite arbitrary. Completely general solutions for the latent probability models have been obtained, handling items with any number of manifest response categories over any number of time points. Here we shall consider only the case of items with three response categories, where there are also three latent classes, and where the data are collected for three time points or panel waves. It will be evident by induction that solutions of this type can be extended to many categories and many time points.

In this model change is allowed among all latent classes, without restriction. Aside from the general assumptions of all these latent probability models, we assume only that if the latent position at time 2 is fixed, the latent position at time 3 does not depend on the latent position at time 1. This is the same assumption of independence of the latent initial and terminal positions which was used for the dichotomous cases in the text.

Let the subscript "1" stand for the first category, "2" for the second, and "3" for the third. The position of a subscript will indicate the interview to which it refers. These subscripts will be used with "p" for the manifest data and with "v" for the latent proportions. Let a_{ij} be the probability that a respondent in latent class C_i is reported in category j, i.e., is in p_j.

The following equation results from the fundamental assumptions:

$$\begin{vmatrix} 1 & p_{..1} & p_{..2} \\ p_{1..} & p_{1.1} & p_{1.2} \\ p_{2..} & p_{2.1} & p_{2.2} \end{vmatrix} = \begin{vmatrix} 1 & 1 & 1 \\ a_{11} & a_{21} & a_{31} \\ a_{12} & a_{22} & a_{32} \end{vmatrix} \tag{1}$$

$$\cdot \left| \begin{pmatrix} v_{111} & v_{112} & v_{113} & v_{121} & v_{122} & v_{123} & v_{131} & v_{132} & v_{133} \\ v_{211} & v_{212} & v_{213} & v_{221} & v_{222} & v_{223} & v_{231} & v_{232} & v_{233} \\ v_{311} & v_{312} & v_{313} & v_{321} & v_{322} & v_{323} & v_{331} & v_{332} & v_{333} \end{pmatrix} \begin{pmatrix} 1 & a_{11} & a_{12} \\ 1 & a_{21} & a_{22} \\ 1 & a_{31} & a_{32} \\ 1 & a_{11} & a_{12} \\ 1 & a_{21} & a_{22} \\ 1 & a_{31} & a_{32} \\ 1 & a_{11} & a_{12} \\ 1 & a_{21} & a_{22} \\ 1 & a_{31} & a_{32} \end{pmatrix} \right|$$

Let us call the first determinant on the right hand side of (1) $|\bar{A}'|$. Call the first rectangular matrix within the second determinant on the right hand side V, and the second matrix within this determinant A. By the theorem of multiplication of determinantal arrays the second determinant is equal to the sum of $\frac{9 \cdot 8 \cdot 7}{1 \cdot 2 \cdot 3} = 84$ products made by pairing each minor of order 3 from 3 columns of V with the minor of order 3 from the corresponding rows of A. The minors from 3 columns of V which enter into these products will be of the following type:

$$\begin{vmatrix} v_{1hi} & v_{1jk} & v_{1rs} \\ v_{2hi} & v_{2jk} & v_{2rs} \\ v_{3hi} & v_{3jk} & v_{3rs} \end{vmatrix} \quad \text{where } h, i \neq j, k \neq r, s. \tag{2}$$

Now let us consider what restrictions the assumption that the latent position at time 3 is independent of the latent position at time 1 given the latent position at time 2 requires. It implies equalities of this kind:

$$\frac{v_{1i1}}{v_{1i2}} = \frac{v_{2i1}}{v_{2i2}} = \frac{v_{3i1}}{v_{3i2}} \tag{3}$$

and

$$\frac{v_{1i1}}{v_{1i3}} = \frac{v_{2i1}}{v_{2i3}} = \frac{v_{3i1}}{v_{3i3}}$$

That is, when the position at time 2 is fixed, the ratio of those going to any given position at time 3 to those going to any other given position at time 3 is the same for all positions at time 1. Now consider a determinant of the following type:

$$\begin{vmatrix} v_{1i1} & v_{1i2} & v_{1j3} \\ v_{2i1} & v_{2i2} & v_{2j3} \\ v_{3i1} & v_{3i2} & v_{3j3} \end{vmatrix} \quad (4)$$

All the elements of two columns of this determinant (we have here chosen the first and second columns for illustration) have the same second subscript. All elements of the remaining column have the same second subscript, which may be the same as or different from that of the other two columns. The first and third subscripts are arranged in the customary manner for determinants.

Expand (4) by the minors of the two columns which have the same second subscript. By (3) all these minors are zero. Therefore (4) vanishes. Hence in the expansion of the determinant of the product of rectangular matrices in (1) all products of paired minors which have minors of V with at least two columns having the same second subscript vanish.

Now stratify the matrix whose determinant is the left hand side of (1) by position 1 at time 2. The determinant of this stratified matrix is then, by the fundamental assumptions of these models, as follows:

$$\begin{vmatrix} p_{.1.} & p_{.11} & p_{.12} \\ p_{11.} & p_{111} & p_{112} \\ p_{21.} & p_{211} & p_{212} \end{vmatrix} = \tag{5}$$

$$|\bar{A}'| \left| \begin{pmatrix} v_{111} & v_{112} & v_{113} & v_{121} & v_{122} & v_{123} & v_{131} & v_{132} & v_{133} \\ v_{211} & v_{212} & v_{213} & v_{221} & v_{222} & v_{223} & v_{231} & v_{232} & v_{233} \\ v_{311} & v_{312} & v_{313} & v_{321} & v_{322} & v_{323} & v_{331} & v_{332} & v_{333} \end{pmatrix} \begin{pmatrix} a_{11} & a_{11}a_{11} & a_{11}a_{12} \\ a_{11} & a_{11}a_{21} & a_{11}a_{22} \\ a_{11} & a_{11}a_{31} & a_{11}a_{32} \\ a_{21} & a_{21}a_{11} & a_{21}a_{12} \\ a_{21} & a_{21}a_{21} & a_{21}a_{22} \\ a_{21} & a_{21}a_{31} & a_{21}a_{32} \\ a_{31} & a_{31}a_{11} & a_{31}a_{12} \\ a_{31} & a_{31}a_{21} & a_{31}a_{22} \\ a_{31} & a_{31}a_{31} & a_{31}a_{32} \end{pmatrix} \right|$$

All the rows of A which correspond to columns of V having the second subscript 1 have been multiplied by a_{11}; all rows of A corresponding to columns of V having the second subscript 2 have been multiplied by a_{21}; and all rows of A corresponding to columns of V having the second subscript 3 have been multiplied by a_{31}. Let us call this new matrix A_{111}. By the independence assumption all products in the expansion of VA_{111} which have two rows multiplied by the same a_{i1} will vanish, because the minor from V in such products will vanish. The remaining terms will be products of minors of V and minors of A with one of the three rows of the minor of A multiplied by a_{11}, one by a_{21}, and one by a_{31}. The a_{i1} can be factored out, and the remaining expression is the expansion of VA. Hence the left hand side of (5) is equal to

$$a_{11}a_{21}a_{31}|\bar{A}'|\,|VA|$$

Dividing (5) by (1)

$$\frac{\begin{vmatrix} p_{.1.} & p_{.11} & p_{.12} \\ p_{11.} & p_{111} & p_{112} \\ p_{21.} & p_{211} & p_{212} \end{vmatrix}}{\begin{vmatrix} 1 & p_{..1} & p_{..2} \\ p_{1..} & p_{1.1} & p_{1.2} \\ p_{2..} & p_{2.1} & p_{2.2} \end{vmatrix}} = a_{11}a_{21}a_{31} \tag{6}$$

Now form a determinant by substituting one of the columns of manifest data, say the first column, from the stratified determinant of (5) for the corresponding column of manifest data from the unstratified determinant of (1). This results in the following equation:

$$\begin{vmatrix} p_{.1.} & p_{..1} & p_{..2} \\ p_{11.} & p_{1.1} & p_{1.2} \\ p_{21.} & p_{2.1} & p_{2.2} \end{vmatrix} = |\bar{A}'| \left| V \begin{pmatrix} a_{11} & a_{11} & a_{12} \\ a_{11} & a_{21} & a_{22} \\ a_{11} & a_{31} & a_{32} \\ a_{21} & a_{11} & a_{12} \\ a_{21} & a_{21} & a_{22} \\ a_{21} & a_{31} & a_{32} \\ a_{31} & a_{11} & a_{12} \\ a_{31} & a_{21} & a_{22} \\ a_{31} & a_{31} & a_{32} \end{pmatrix} \right| \qquad (7)$$

The equation is the same as (1), except that the elements of the first column of A have been multiplied by some a_{i1}, where i for a given row of A is the same as the second subscript of the corresponding column of V. By the same reasoning previously used, it is clear that in the expansion of the determinant of the product of V and this new A matrix, all the terms vanish which have any given a_{i1} as the first element in two rows of the minor taken from the modified A matrix. The remaining terms all have, in the minors taken from the modified A matrix, a_{11} as the first element of one row, a_{21} as the first element of another row, and a_{31} as the first element of the third row. Only in the first column do these minors differ from the corresponding minors of A in the expansion of $|VA|$.

We next form a determinant by substituting the second column from the stratified determinant for the corresponding column of the unstratified determinant of (1). When this is put in terms of $|\bar{A}'|$, V, and the necessary modification of A, and when the determinant of the product of V and this modified A is expanded, the terms are identical with those of the previous expansion, except that now it is the second column of each minor taken from A which has one element multiplied by a_{11}, one by a_{21}, and one by a_{31}. When the third column of the stratified determinant is substituted for the third column of (1), it is the third column of each minor from A which has one element multiplied by a_{11}, one by a_{21}, and one by a_{31}. Further, the corresponding terms of the three expansions have, in the minors taken from A, the same row (but different columns) multiplied by a_{11}, a_{21} or a_{31}, as the case may be. Add the expansions of the three determinants which have been formed by substituting one column of (5) for the corresponding column of (1), and sum with respect to the corresponding terms, which are identical in minors taken from V. The result is the product of $|\bar{A}'|$ and a sum of terms of the following type:

$$\begin{vmatrix} v_{1hi} & v_{1jk} & v_{1rs} \\ v_{2hi} & v_{2jk} & v_{2rs} \\ v_{3hi} & v_{3jk} & v_{3rs} \end{vmatrix} \tag{8}$$

$$\left\{\begin{vmatrix} a_{h1} & a_{i1} & a_{i2} \\ a_{j1} & a_{k1} & a_{k2} \\ a_{r1} & a_{s1} & a_{s2} \end{vmatrix} + \begin{vmatrix} 1 & a_{h1}a_{i1} & a_{i2} \\ 1 & a_{j1}a_{k1} & a_{k2} \\ 1 & a_{r1}a_{s1} & a_{s2} \end{vmatrix} + \begin{vmatrix} 1 & a_{i1} & a_{h1}a_{i2} \\ 1 & a_{k1} & a_{j1}a_{k2} \\ 1 & a_{s1} & a_{r1}a_{s2} \end{vmatrix}\right\}$$

where $h \neq j \neq r$.

Expanding the three determinants in brackets by the first, second, and third columns respectively gives

$$a_{h1}\begin{vmatrix} a_{k1} & a_{k2} \\ a_{s1} & a_{s2} \end{vmatrix} - a_{j1}\begin{vmatrix} a_{i1} & a_{i2} \\ a_{s1} & a_{s2} \end{vmatrix} + a_{r1}\begin{vmatrix} a_{i1} & a_{i2} \\ a_{k1} & a_{k2} \end{vmatrix}$$

$$-a_{h1}a_{i1}\begin{vmatrix} 1 & a_{k2} \\ 1 & a_{s2} \end{vmatrix} + a_{j1}a_{k1}\begin{vmatrix} 1 & a_{i2} \\ 1 & a_{s2} \end{vmatrix} - a_{r1}a_{s1}\begin{vmatrix} 1 & a_{i2} \\ 1 & a_{k2} \end{vmatrix}$$

$$+a_{h1}a_{i2}\begin{vmatrix} 1 & a_{k1} \\ 1 & a_{s1} \end{vmatrix} - a_{j1}a_{k2}\begin{vmatrix} 1 & a_{i1} \\ 1 & a_{s1} \end{vmatrix} + a_{r1}a_{s2}\begin{vmatrix} 1 & a_{i1} \\ 1 & a_{k1} \end{vmatrix}$$

Now the sum of the three factors by which a_{h1} is multiplied is nothing but an expansion of a minor of A by its first row; the sum of the three factors multiplying a_{j1} is an expansion of the same minor by its second row; and the sum of three factors multiplying a_{r1} is an expansion of the same minor by its third row. Hence the expression in brackets of (8) equals

$$(a_{h1} + a_{j1} + a_{r1})\begin{vmatrix} 1 & a_{i1} & a_{i2} \\ 1 & a_{k1} & a_{k2} \\ 1 & a_{s1} & a_{s2} \end{vmatrix}$$

But since $h \neq j \neq r$,

$$a_{h1} + a_{j1} + a_{r1} = a_{11} + a_{21} + a_{31}$$

for any h, j, and r. Therefore this sum can be factored out of the sum of products of paired determinantal expressions. What remains is merely the expansion of $|VA|$. Hence, finally,

$$\begin{vmatrix} p_{.1.} & p_{..1} & p_{..2} \\ p_{11.} & p_{1.1} & p_{1.2} \\ p_{21.} & p_{2.1} & p_{2.2} \end{vmatrix} + \begin{vmatrix} 1 & p_{.11} & p_{..2} \\ p_{1..} & p_{111} & p_{1.2} \\ p_{2..} & p_{211} & p_{2.2} \end{vmatrix} + \begin{vmatrix} 1 & p_{..1} & p_{.12} \\ p_{1..} & p_{1.1} & p_{112} \\ p_{2..} & p_{2.1} & p_{2.2} \end{vmatrix} \quad (9)$$

$$= (a_{11} + a_{21} + a_{31})|\bar{A}'||VA|$$

Evidently (9) divided by (1) equals $a_{11} + a_{21} + a_{31}$.

Next take the determinant formed by substituting the first two columns of the stratified determinant (5) for the first two columns of the unstratified determinant (1). This determinant will equal the product of $|\bar{A}'|$ and the determinant of the product of V and a matrix which is A with the first two elements of each row multiplied by some a_{i1}, where i in a given row is the same as the second subscript in the corresponding column of V. Take also the other two determinants which can be formed by substituting two columns of the stratified determinant for the corresponding columns of the unstratified determinant (1). Sum the three determinants formed in this way. When the sum is evaluated, it may be shown by the same reasoning employed above that it equals

$$(a_{11}a_{21} + a_{11}a_{31} + a_{21}a_{31})|\bar{A}'||VA|$$

Dividing this by (1) gives $a_{11}a_{21} + a_{11}a_{31} + a_{21}a_{31}$.

Let us call the unstratified determinant which is the left hand side of (1) $|P|$, and the stratified determinant which is the left hand side of (5) $|P_{111}|$. Call the sum of determinants formed from two columns of $|P|$ and one of $|P_{111}|$ $|P_1|$. Call the sum of the determinants formed from one column of $|P|$ and two of $|P_{111}|$ $|P_{11}|$. Then a_{11}, a_{21}, and a_{31} can be found by solving the cubic equation

$$x^3 - \frac{|P_1|}{|P|}x^2 + \frac{|P_{11}|}{|P|}x - \frac{|P_{111}|}{|P|} = 0 \quad (10)$$

a_{12}, a_{22}, and a_{32} can be obtained in exactly the same manner by using $|P|$ and the stratification of $|P|$ by position 2 at time 2. Similarly, the a_{i3} can be obtained by using $|P|$ and the stratification by position 3 at time 2. It might be thought unnecessary to use position 3, since from the equations

$$\sum_j a_{ij} = 1$$

a_{i3} could be determined. But we have not yet identified the roots of (10) and the corresponding equation in a_{i2}; that is, we do not know which root of (10), for example is a_{11}, which is a_{21}, and which is a_{31}. Hence we have no way of pairing the roots of (10) and the roots of the corresponding equation in a_{i2} in in order to subtract from 1. In order to identify the roots, we must also solve for the a_{i3}.

The procedure then is as follows. If the item has even moderate reliability for the various classes, $a_{ii} > a_{ji}$, $i \neq j$, for any i and j. Hence from any set of roots, say a_{i1}, select the largest root and call it a_{11}. Another way of describing this procedure is to say that we have, not altogether arbitrarily, designated as latent class C_i that class which has the greatest probability of giving the response i. Having done this, we then make use of the equations

$$\sum_{\substack{j \\ j \neq i}} a_{ij} = 1 - a_{ii}$$

to pair the other roots.

This will give a unique solution in nearly all cases. For suppose that the actual probabilities are $a_{11}, a_{12}, \ldots, a_{33}$. If a solution to the equation above is indeterminate, we must have

$$1 - a_{11} = a_{12} + a_{13} = a_{32} + a_{23}$$

$$1 - a_{22} = a_{21} + a_{23} = a_{31} + a_{13}$$

$$1 - a_{33} = a_{31} + a_{32} = a_{21} + a_{12}$$

from which we get

$$a_{13} - a_{23} = a_{21} - a_{31} = a_{32} - a_{12}$$

Only in cases where this special condition is met will the solution fail to be unique. Even in such cases, intuitive judgments will often provide a satisfactory means of decision.

Once the a_{ii} have been properly identified, it is a simple matter to find the v_{ijk}. Let us define matrices P_{12}, P_{23}, V_{12}, and V_{23} as follows:

$$P_{12} = \begin{pmatrix} 1 & p_{.1.} & p_{.2.} \\ p_{1..} & p_{11.} & p_{12.} \\ p_{2..} & p_{21.} & p_{22.} \end{pmatrix} \qquad P_{23} = \begin{pmatrix} 1 & p_{..1} & p_{..2} \\ p_{.1.} & p_{.11} & p_{.12} \\ p_{.2.} & p_{.21} & p_{.22} \end{pmatrix}$$

$$V_{12} = \begin{pmatrix} v_{11.} & v_{12.} & v_{13.} \\ v_{21.} & v_{22.} & v_{23.} \\ v_{31.} & v_{32.} & v_{33.} \end{pmatrix} \qquad V_{23} = \begin{pmatrix} v_{.11} & v_{.12} & v_{.13} \\ v_{.21} & v_{.22} & v_{.23} \\ v_{.31} & v_{.32} & v_{.33} \end{pmatrix}$$

From the fundamental assumptions,

$$P_{12} = \bar{A}' V_{12} \bar{A}$$

and

$$P_{23} = \bar{A}' V_{23} \bar{A}$$

Since we know $\bar{A}$, we can find $\bar{A}^{-1}$. We can then find the joint true classes for times 1 and 2 and those for times 2 and 3 by

$$V_{12} = \bar{A}'^{-1} P_{12} \bar{A}^{-1} \tag{11}$$

and

$$V_{23} = \bar{A}'^{-1} P_{23} \bar{A}^{-1}$$

From V_{12} and V_{23} by simple addition of rows or columns can be found the proportions in each latent class at times 1, 2, and 3 individually. To find the triple occurrence latent classes requires the independence assumption. This assumption implies that

$$\frac{v_{ijk}}{v_{ij.}} = \frac{v_{hjk}}{v_{hj.}} = \frac{v_{gjk}}{v_{gj.}} = \frac{v_{.jk}}{v_{.j.}}$$

or

$$v_{ijk} = \frac{v_{ij.}\, v_{.jk}}{v_{.j.}}$$

Hence the triple occurrence latent proportions can be determined from the elements of V_{12} and V_{23} and either the column marginals of V_{12} or the row marginals of V_{23}, which give the $v_{.j.}$.

It is immediately apparent that the equations of this section, stated for items with three response categories, are applicable, *mutatis mutandis*, to items with any number of response categories, and to any number of time points.

APPENDIX C

Design of the Study of Level of Aspiration and Performance of Sociology Students

The Task: A Course in Methods of Social Research

The subjects chosen for the study were graduate students in sociology at Columbia University, and the task with reference to which their aspirations and performances were studied was a course in methods of social research. This course was required for all Ph.D. candidates in sociology. A brief description of the course will be given here, for purposes of general orientation, but it must be kept in mind that in this study the content of the course does not figure in any important manner. An interesting study could be done of the content of the course in relation to student interests and aspirations, but such a study would be quite different from the aims of this one. In this case, the content of the course is merely a confounding and extraneous factor, of no more interest in itself than the sort of tasks used in the typical level of aspiration experiment, such as dart-throwing or puzzle-solving. What is important about the nature of the task is that most students felt that achievement in the course was essential to the accomplishment of their career aims, so that they were ego-involved and their strength of motivation was relatively high.

There were, however variations through time in the content of the course, which could impinge significantly on some of the variables under study, such as the perception of task difficulty, interest in the course, or attractiveness of the course, at any particular time. These variations are discounted by the fact that attention is paid only to differences between groups, so that relative rather than absolute levels of variables are the focus of concern. If there is a significant differential change in perception of task difficulty between two groups of subjects, it is immaterial whether the actual difficulty rose or fell. For this reason, we are not here concerned with the change in content of the course.

For general orientation, nonetheless, it is useful to understand the nature of the task under consideration. The course in methods of social research was studied for the two academic years 1948—49, and 1949—50. Since data from the former year alone appear in this paper, the course in 1949—50 will not be described. The course given the former year was a first-year graduate course in methods, which assumed no previous training or experience in social research. It was organized into three sections, each with thirty to forty students, each section meeting once a week for two hours. Two of the sections were taught by an instructor in sociology who had considerable research experience, while the third was taught by an assistant. The instructor and the assistant worked in very close collaboration, insuring that the content of the course at any given time was virtually identical for the three sections. For this reason the three sections have been combined in the analysis. Although there are certain minor differences among the sections, these are almost entirely functions of the different hours at which the section meetings were held, which attracted differentially students who were working and those who were not. Such differences are extraneous in the present study.

As a first-year graduate course in methods of research, the course covered a great variety of topics. Broadly, the first semester was concerned with problems of design, questionnaire construction, choosing a sample, and operational field problems. The second semester dealt mainly with problems of analysis of field data, including coding, content analysis, IBM procedures, index construction, relations of variables, and interpretation of results.

There are, of course, correlations between certain variations in content through time, particularly in the degree of technical difficulty in the content, and the attitudes of certain sub-groups of the students towards the course. Those with technical proclivities or interests generally felt that they were learning more when the course was most technical, while those who were afraid of or hostile to technical matters found the less technical subjects more stimulating.

We are eschewing considerations of content of the task in order to raise the level of abstraction of the analysis to the point where it will mesh with the conceptual framework of level of aspiration as it has been developed in experimental social psychology. Happily, an independent study has been made of the same course which focuses on content, teaching methods, and student reactions to these (see Wright, 1952). The reader interested in supplementing his knowledge of the task which is used here in a rather formal way as an occasion for the study of relationships among psychological and sociological variables is referred to that study.

The formal course procedures and requirements leading to assignment of grades, in contrast to content of the course, are important for a study of grade aspiration and achievement. Students were required to submit homework at irregular intervals, and this homework was graded and returned to the student. In addition, especially during the year 1948—49, when sections were small, students were encouraged to speak up in class and to ask and answer questions. Each student, furthermore, was urged to participate in an actual research project, supervised by some outside person, usually a member of the staff of the Bureau of Applied Social Research at Columbia University, or as a less desirable alternative, to complete a small research project of his own. Those who did engage in an actual research project were not required to take an examination at the end of the term, while those who did not do research had to take such an examination. Only about one-tenth of the students took the final examination. For these, the grade on the examination counted very heavily in the final grade, while for the majority who did not take the examination, the caliber of their contribution to the research projects as judged by the project supervisors was a decisive factor.

Thus the students had certain indications of their performance throughout the term, but of these only the homework, which did not weigh heavily in the final grade, provided a quantitatively precise measure of achievement. The student's estimate of his achievement depended largely on vague or peripheral indications until the actual grade was received at the end of the first semester.

The Research Design

Twelve written questionnaries were administered to members of the methods course during the academic year 1948—49. Ten of these questionnaires constitute the panel proper and the other two are supplementary. The first questionnaire of the panel was administered at the second class meeting in the fall, and subsequent waves were administered at every third class meeting. There were fifteen meetings each semester, not counting examinations, so that altogether five waves were obtained for each semester, or a total of ten for the academic year. The decisive time interval was taken as the number of class meetings intervening between two waves, rather than the calendar time. Where school holidays or examination periods occurred, the calendar time between waves was lengthened by one or two weeks. The effect of such vacations was to spread the ten waves over a period from the beginning of October to the end of May, yielding an average of almost a month of calendar time between waves instead of three weeks. The successive administrations of questionnaires in the panel proper will be referred to as "Wave 1", "Wave 2", etc. Since two waves were administered in October and in March, slight variants in designation are required to specify these waves. The actual dates of administration and the designations which were used are shown in Table 43.

The core of the questionnaires, from the point of view of studying changes through time, is a set of seven questions which were repeated on a number of waves. These questions were designed to tap seven interlocking psychological variables. It is worth noting here that the use of single items as indices of psychological variables involves a certain crudity which is the price that had to be

TABLE 43

Terminology and administration dates of panel waves, 1948—49

Wave number	*Dates of administration*	*Designation*
1	October 4—7	Initial wave
2	October 25—28	October
3	November 15—18	November
4	December 13—16	December
5	January 10—13	January
6	February 7—10	February
7	February 28—March 3	Early March
8	March 21—24	Late March
9	April 18—21	April
10	May 9—12	May

paid for employing ten waves. It was not possible within the limited time available for administration to use batteries of items which could then be transformed into scales. Instead, graphic self-rating scales were employed in most of the later waves.

Unfortunately, the entire set of seven items was not employed on all ten waves. Only four of the items were employed on the first three waves, six of the items were used on the fourth wave, and all seven items were used on the fifth wave and all successive waves. At the beginning, it was not evident that all seven items would be required, but preliminary analysis of earlier waves showed the need for additional information, which was then incorporated into later waves. The possibility of such additions, of course, is one of the virtues of the panel design. One of the questionnares which includes the variables referred to in this paper may be found at the end of this Appendix.

In addition to the repeated items, various waves of the panel included other items which were not usually repeated but provided information required for the analysis. The most important wave in this respect is the February wave, or wave 6, which came at the beginning of the second semester. Prior to that time, as noted above, the students had relatively little objective indication of their actual achievement in the course. Such indications as there were were based on their homework assignments which had been returned to them throughout the first term, and on their contacts with the teacher, with their project directors, and with other students. Between the January and February waves, however, grades were assigned to nearly all the students (a few received the grade of "Incomplete"). Since the reactions of subjects to the grades they received is one of the major interests in such a study, items were included in the February and early March questionnaires to measure these reactions directly.

In addition to the ten panel waves, two other questionnaires were administered to the class. The first of these, administered in January, contained a set of items on background characteristics of the student, such as age, sex, and family income, which it was not convenient to include in the panel questionnaires. The second additional instrument contained a series of questions about students' reactions to the course and to the panel questionnaires with emphasis on the subjective definition by the student of variables used in the study. Students were asked by the teacher to submit answers to these questions as a homework assignment. The assignment was handed out at the time of the early March wave, but was completed by students at varying times throughout the remainder of the semester.

In summary, the material available for the year 1948—49 includes ten panel waves with a set of repeated items, supplementary data taken on various waves, especially on waves immediately following the grade assignment, additional background information, and qualitative data on the students' attitudes towards the course and their interpretations of the repeated questions used in the panel.

In the administration of the panel questionnaires, students were given on the initial wave a rationale for the study which emphasized the following points:

(1) The study has sociological value.

(2) Responses to questions will have no influence on the student's grade in the course, since the teacher of the course is not connected with the study and will not see the questionnaires.

(3) The student is guaranteed anonymity by using a code number assigned at random rather than his name as identification of his protocols.

Portions of this rationale were repeated at the administration of subsequent waves, and the cooperation obtained from students was generally quite good. On the homework assignment later students were asked whether they believed that their questionnaires were actually anonymous, and whether they believed their responses had any effect on their grades. Most students believed that anonymity was preserved, and most of those who did not, felt that it made no difference as far as their grades or the attitudes of the teacher were concerned.

Needless to say, a procedure was employed which permitted identification of the students, since their actual grades had to be collated with their questionnaires. However, the questionnaires were never made available to the teacher, and the study had no effect on the teachers' estimates of the students.

Sample Size, Mortality, and Usable Sample

It is not quite accurate to refer to the "sample size" in this study, because strictly speaking, a random sample is not involved. For convenience, the term will be used with this reservation. There are many complications about sample size in panel studies, mainly resulting from the "mortality" or missing responses on particular waves (see Glock, 1952). In the present study, the total possible number of cases is the number of students who enrolled for the course in research methods. Even this is complicated by the fact that some students took the course the first semester and not the second, and some who had not taken the first semester enrolled for the second. The numbers for the year 1948—49 are shown in Table 44.

TABLE 44

Enrollment in methods course, 1948—49

	Winter semester	*Spring semester*	*Both semesters*
1948—49	118	123	90

For the study of relationships at any one time, the possible maximum number of cases is the number enrolled at that time, but for studying changes through time which overlap both semesters, the possible maximum is the number of students enrolled for both semesters.

There are several reasons why the number of respondents on any given panel wave falls short of the total number enrolled. In addition to the lack of cooperation, there is the fact that students missed class sessions because of illness, lack of interest, or other reasons, and on occasion students came to class late, had to leave early, and so on. During the first year a considerable effort was made to pick up these missing cases by "call backs", that is, by administering the questionnaire the following week, or by locating the student outside of class and having him fill it out. While this might have introduced a slight error into the data (because the student's attitudes might have changed a little between the time of regular administration and the time of "call-back"), it was probably less serious than the loss of the case.

The total number of completed questionnaires for each panel wave and for the additional questionnaires is shown in Table 45.

These figures show the actual number of cases available for analysis of any given wave, but they are at the same time too small to show the total extent of participation in the study and too large if they are taken as indicating the number of cases available for analysis of change or latent process.

A more adequate picture is obtained by looking at the number of respondents who completed questionnaires on all ten waves, on

TABLE 45

Number of respondents, all questionnaires

	Panel waves										*Background*
	1	*2*	*3*	*4*	*5*	*6*	*7*	*8*	*9*	*10*	
1948–49	101	96	93	99	100	121	120	112	110	91	132

nine waves, eight waves, and so on. This is the joint distribution of participation over all panel waves, in other words, and it is shown in Table 46.

Once the distributions are given, serious problems appear. A major aim of this study is to follow some attitude or perception of a given group of students from beginning to end of the school year. If a given student has responded to all the panel waves, there is no problem. But what if he has responded to six waves, say, in the first year? We have substantial data on such a person, yet for four waves we have no information.

Suppose, to be more concrete, we want to consider the average level of interest in the course of students who made a particular grade, and then compute the average at each time, simply averaging for the sub-set of respondents who happened to complete the questionnaire at that time. The number of cases and the identity of some of the respondents would vary from wave to wave. In this way we would use all the available data for those who fell into the

TABLE 46

Joint distribution of respondents over all panel waves

No. waves completed	*1948–49*
10	40
9	24
8	10
7	8
6	8
5	31
4	14
3	8
2	4
1	4
Total No. participants	151

grade group in question, and we would keep the number of cases at a maximum.

To use this procedure, however, would involve a serious error. The students who fail to respond on a given panel wave are not a random sample of the class; they differ systematically in many ways from those who do respond (Glock, 1952). In the present example, the average interest in the course of students who miss a panel wave is lower than that of those who respond, even among any particular grade-group. Thus computing average interest omitting these cases will give a higher value than would be obtained had they responded.

To make matters worse, the mortality is by no means constant at each time, as shown above. Suppose that in a given group, everyone responds to wave 3. On wave 4, just before Christmas, students with lower interest in the course go home on vacation early and miss the lecture. Computing the average for those who remain will make it appear that average interest among this group has risen, even though in fact it may not have changed or may have even decreased. The absence of a biased sub-set of respondents will distort the picture of change, usually in an unknown manner. It may also distort correlations between variables by altering the range of one or both of the variables.

Considerations of this sort have led to a basic principle in the handling of cases, which is adhered to (with one exception) throughout whenever time series or comparisons over time are involved; a respondent is either excluded altogether or he is included at every point. If he is excluded, a bias may occur in findings at any one time; but we are not dealing with a random sample in the first place, and more important, the bias will not alter from one time to the next. Comparisons through time, whether of averages or correlations, will not be distorted.

If, on the other hand, a respondent is not excluded, the problem remains how to deal with the waves which he has missed. For a number of reasons, some method of interpolating values for the missing waves seems to be the best solution. Basically, interpolation yields less distortion than omitting a value because the order of magnitude of differences among persons is generally greater than the order of magnitude of differences within the same person over time. As an empirical fact, people differ from each other more on almost any variable than the average person differs with himself at different times. Thus omitting a biased sub-set of values creates a greater error than interpolating values on the basis of the values at waves immediately before and after the missing wave.

A hypothetical example but one typical of the data will illustrate this point. Values on the variable "interest in the course" for three cases on three waves might be as follows:

	Wave 1	*Wave 2*	*Wave 3*
Case 1	90	80	90
Case 2	70	80	80
Case 3	50	missing	50
Average without interpolation	70	80	73
Average with interpolation (50)	70	70	73

Taking an average at wave 2 without interpolation gives a mean of 80, while interpolating a value of 50 for the missing case yield a mean of 70. If case 3 had responded on wave 2, he would be much more likely to give a value of 50, the mean of his own values at waves 1 and 3, than a value of 80, the mean of case 1 and case 2 at wave 2. Hence the interpolation yields a better estimate of the mean at wave 2 than the omission of case 3.

Many methods of interpolation could be devised, but the simplest method, linear interpolation, seemed also to be the best. Where the first or last wave was missing, the value of the variable on the second or ninth wave was used.

Clearly, interpolation of values in this manner has certain disadvantages. The principal artifact which it introduces occurs when the same variable is correlated with itself at a successive wave. In that case, the interpolation of values on, say, the second wave based on values obtained on the first will tend to inflate the correlation spuriously. Put another way, the effect of linear interpolation is to smooth out trends or to minimize changes. But in this study, the great majority of cases show little change from one wave to the next on any particular variable. The effect of linear interpolation is usually to attenuate but rarely to exaggerate findings of change.

On the other hand, excessive attenuation or smoothing out of changes is certainly undesirable. Obviously the less interpolation which has to be done the smaller will be the consequent distortion. We are caught here between the desire to reduce error by eliminating from the analysis as few cases as possible and the desire to reduce error by interpolating as little as possible. Some more or less arbitrary rule of thumb must be employed to decide which respondents to include and which to exclude.

The rule adopted for the 1948—49 panel was to include only respondents who completed seven or more waves of the panel, and who did not miss more than two waves in succession at any time during the year. The only exception to this was the inclusion of five respondents who missed the first three waves but completed the last seven. These respondents were absent at the first wave, and failed to complete the second and third solely through a misunderstanding. Because their lack of response to the first three waves was a result of misunderstanding rather than a biasing factor, they were included in the panel analysis. Values were not interpolated for the first three waves for these five respondents, however, thus making them also the only exception to the general principle stated above, namely, that every respondent is either excluded altogether or included at every time point.

With the exception noted, every respondent included in any time series of the 1948—49 class has completed seven or more waves of the panel, and has never missed more than two waves in succession. Application of this rule left a total of 81 cases. Out of this total, ten cases were eliminated because their actual grades were not available, information necessary for substantive findings in other analysis of the data. Thus, the correlations of variables between waves referred to in this monograph are based on a total of 71 cases.

PANEL QUESTIONNAIRE — 3, 4, 5

SOCIOLOGY 195

DON'T TRY TO BE CONSISTENT WITH PREVIOUS REPLIES STATE YOUR PRESENT OPINIONS

(1) Mark the following scale to show how difficult you find the work in this course *at the present time*:

very difficult	rather difficult	average	rather easy	very easy

(2) Indicate how important it is to you to make a good showing on this course by marking the following scale in the appropriate place:

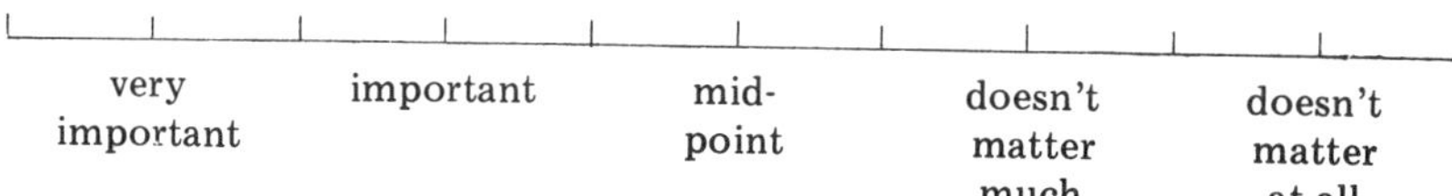

(3) There are seven possible grades for this course — A, A–, B+, B, B–, C+, and C. You and everyone else in the class would of course be delighted to make an A. But which of the seven possible grades would you actually like to attain in Sociology 195; that is, what would you consider a "good" grade in this course, not in general, but for yourself, with your own particular interests and objectives?

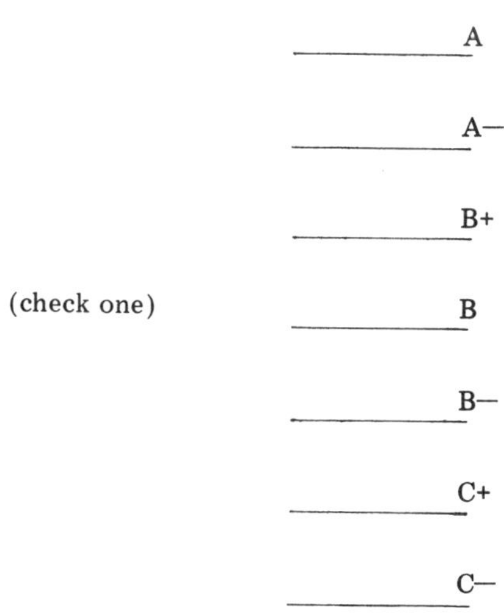

(4) As distinct from your interest in this course, which may be due in part to various extrinsic factors, such as the kind of job you expect to get, how much do you like the course; that is, how much intrinsic enjoyment are you getting out of it *at the present time*? Indicate on the scale, by making a mark on it. A score of +100 on the scale means maximum liking; a score of –100 means maximum disliking.

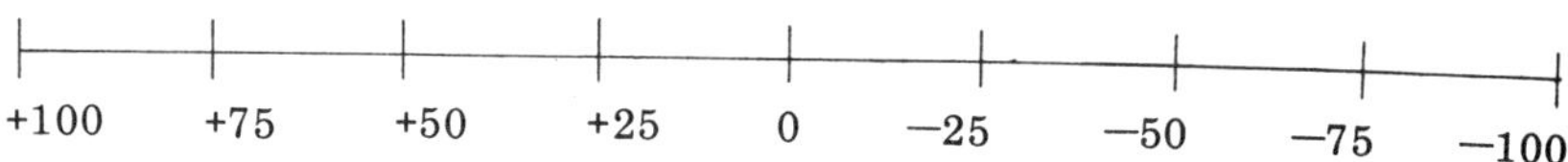

(5) Without reference to grades you have made on other courses, and solely on the basis of your experience with this course, what grade to you think realistically you will probably get on this course?

	________	A
	________	A—
	________	B+
(check one)	________	B
	________	B—
	________	C+
	________	C—

(6) Mark the following scale to indicate what you feel has been the success of your performance on this course during the past month:

very successful	moderately successful	neither	rather unsuccessful	very unsuccessful

(7) How valuable do you think the *present* content of this course is to you?

great value	worth-while	possible value	little value	worth-less

References

Aitken, A.C. (1948). *Determinants and Matrices.* 5th ed. New York: Wiley Interscience Publishers.

Anderson, T.W. (1951). "Probability models for analyzing time changes in attitudes." Part VI of Lazarsfeld, P.F., ed., *The Use of Mathematical Models in the Measurement of Attitudes.* Santa Monica: The RAND Corporation.

Anderson, T.W. (1954). "Probability models for analyzing time changes in attitudes." In Lazarsfeld, P.F., ed., *Mathematical Thinking in the Social Sciences.* Glencoe: The Free Press.

Barlow, R.E., Proschan, F. and Hunter, L.C. (1965). *Mathematical Theory of Reliability.* New York: John Wiley and Sons.

Berelson, B.R., Lazarsfeld, P.F. and McPhee, W.N. (1954). *Voting.* Chicago: The University Press.

Blalock, H.M. (1960). *Social Statistics.* New York: McGraw-Hill.

Blumen, I., Kogan, M. and McCarthy, P.J. (1955). *The Industrial Mobility of Labor as a Probability Process.* Ithaca, N.Y.: Cornell University Press.

Boudon, R. (1967). *L'analyse Mathematique des Faits Sociaux.* Paris: Plan.

Boudon, R. (1968). "A new look at correlation analysis." In Blalock, H.M. and Blalock, A.B., ed., *Methodology in Social Research.* New York: McGraw-Hill.

Campbell, D.T. and Stanley, J.C. (1963). "Experimental and quasi-experimental designs for research and teaching." In Gage, N.L., ed., *Handbook of Research in Teaching.* Chicago: Rand McNally.

Carnap, R. (1936, 1937). "Testability and meaning," *Philosophy of Science,* Vols. III and IV.

Coleman, J.S. (1964). *Models of Change and Response Uncertainty.* Englewood Cliffs: Prentice-Hall.

Cramèr, H. (1946). *Mathematical Methods of Statistics.* Princeton: Princeton University Press.

Cronbach, L.J. (1947). "Test reliability: its meaning and determination," *Psychometrika,* 12: 1—16.

"Evaluation of Project ENABLE." (1967). Mimeo. On deposit in the library of the Family Service Association of America, New York.

Feller, W. (1957). *An Introduction to Probability Theory and Its Applications*, 2nd ed. New York: John Wiley and Sons.

Ferber, R. *et al.* (1959—1966). *Studies in Consumer Savings*, Vols. 1—6. Urbana: Bureau of Economic and Business Research, University of Illinois.

Festinger, L. (1957). *A Theory of Cognitive Dissonance*. Stanford: Stanford University Press.

Frankel, L.R. (1961). "Latent behavior functions." In *Look Audience Study*. New York: Look Magazine.

Glock, C.Y. (1952). Participation Bias and Re-Interview Effect in Panel Studies. Unpublished Ph.D. dissertation, Columbia University. Ann Arbor: University Microfilms.

Gulliksen, H. (1950). *Theory of Mental Tests*. New York: Wiley and Sons.

Guttman, L. (1954). "A new approach to factor analysis: the radex." In Lazarsfeld, P.F., ed., *Mathematical Thinking in the Social Sciences*. Glencoe: The Free Press.

Hart, A.G. (1966). "Arrival at workable stage of Consumer Union panel archive." In Consumer Union Panel Codebook. Mimeo. New York: Columbia University.

Hovland, C.T., Lumsdaine, A.A., and Sheffield, F.D. (1949). *Experiments on Mass Communication*. Princeton: Princeton University Press.

Juster, F.T. (1964). *Anticipations and Purchases: An Analysis of Consumer Behavior*. Princeton: Princeton University Press.

Kendall, P.L. (1954). *Conflict and Mood*. New York: The Free Press.

Kiesler, C.A., Collins, B.E. and Miller, N. (1969). *Attitude Change*. New York: John Wiley and Sons.

Kuehn, A.A. (1958). "An Analysis of the Dynamics of Consumer Behavior and Its Implications for Marketing Management." Unpublished Ph.D. Dissertation, Carnegie Institute of Technology, Pittsburgh.

Lazarsfeld, P.F. (1950). "The interpretation and computation of some latent structures." In Stouffer, S.A. *et al.*, *Measurement and Prediction*, Vol. IV of *Studies in Social Psychology in World War II*. Princeton: Princeton University Press.

Lazarsfeld, P.F. (1954). "A conceptual introduction to latent structure analysis." In Lazarsfeld, P.F., ed., *Mathematical Thinking in the Social Sciences*. Glencoe: The Free Press.

Lazarsfeld, P.F. and Dudman, J. (1951). "Mathematical developments in latent structure analysis". Part II of Lazarsfeld, P.F., ed., *The Use of Mathematical Models in Measurement of Attitudes*. Santa Monica, Calif.: The RAND Corporation.

Lazarsfeld, P.F. and Henry, N. (1968). *Latent Structure Analysis*. Boston: Houghton Mifflin.

Lazarsfeld, P.F., Berelson, B.R. and Gaudet, H. (1948). *The People's Choice*. New York: Columbia University Press.

Levenson, B. and Wiggins, L.M. (1954). "Notes on impact analysis." In Wiggins, L.M., ed., Dartmouth Seminar on Social Process. Mimeo. In Library of Center for Advanced Study in the Behavioral Sciences (sponsor of Seminar directed by Lazarsfeld, P.F.), Palo Alto and Bureau of Applied Social Research, Columbia University, New York.

Lewin, K. (1935). *A Dynamic Theory of Personality.* New York: McGraw-Hill.
Lewin, K. (1936). *Principles of Topological Psychology.* New York: McGraw-Hill.
McFarland, D.D. (1968). "The Use of Orthogonal Base Vectors to Analyze Relationships Between Quantitative Variables in a Panel Study." Mimeo.
McNemar, Q. (1946). "Opinion-attitude Methodology." *Psychological Bulletin,* 43, No. 4.
Nagel, E. (1939). *Principles of the Theory of Probability.* Monograph, Vol. 1, No. 6 in series "International Encyclopedia of Unified Science." Chicago: University of Chicago Press.
Nagel, E. (1961). *The Structure of Science.* New York: Harcourt, Brace and World.
Pareto, V. (1963). *A Treatise on General Sociology.* Transl. Bongiorno, A. and Livingston, A. with advice and active cooperation of Rogers, J.H., ed., Livingston, A. New York: Dover Publications.
Selvin, H. and Simmel, A. (1953). "The Effects of Oil Progress Week, 1952: A Supplementary Analysis." Mimeo. New York: Bureau of Applied Social Research, Columbia University.
Social Casework. (1967). *Project ENABLE,* Vol. XLVIII, No. 10. New York: Family Service Association of America.
Thurstone, L.L. (1949). *Multiple Factor Analysis.* Chicago: University of Chicago Press.
Tolman, E.C. (1932). *Purposive Behavior in Animals and Men.* New York: The Century Co.
Torgersen, W.S. (1958). *Theory and Methods of Scaling.* New York: John Wiley and Sons.
Wiggins, L.M. (1941). *Reason, Will and Responsibility.* Chapel Hill: University of North Carolina.
Wiggins, L.M. (1951). "The application of latent structure analysis to the problem of reliability". Part III of Lazarsfeld, P.F., ed., *The Use of Mathematical Models in the Measurement of Attitudes.* Santa Monica: The RAND Corporation.
Wiggins, L.M. (1954a). "Foci of interest in the discussion of social process." In Wiggins, L.M., ed., Darmouth Seminar on Social Process. Mimeo. In Library of Center for Advanced Study in the Behavioral Sciences (sponsor of Seminar directed by Lazarsfeld, P.F.), Palo Alto, and Bureau of Applied Social Research, Columbia University, New York.
Wiggins, L.M. (1954b). "Notes on models for a common type of attitude process," as in Wiggins (1954a).
Wiggins, L.M. (1954c). "The multi-wave panel as a quasi-experiment," as in Wiggins (1954a).
Wiggins, L.M. (1955a). "Mathematical Models for the Analysis of Multi-wave Panels." Ph.D. Dissertation, Columbia University. Ann Arbor: University Microfilms.
Wiggins, L.M. (1955b). "A Panel Study of Automobile Buying." Mimeo. Palo Alto: Center for Advanced Study in the Behavioral Sciences.
Wiggins, L.M. (1958). "The Stability of Automobile Images." Mimeo. Bureau of Applied Social Research, Columbia University, New York.

Wiggins, L.M. (1966). "The use of simulation to relate behavioral theory to social action." In Calabro, N., ed., *Proceedings on Simulation in Business and Public Health.* New York: American Statistical Association.

Wright, C. (1952). "A Qualitative Analysis of Factors Affecting the Training of Students for Modern Social Research." Mimeo. Report prepared for Planning Project for Advanced Training in Social Research. Bureau of Applied Social Research, Columbia University, New York.

Index